Analytical Methods in Physics

Luiza Angheluta

Analytical Methods in Physics

 Springer

Luiza Angheluta
Department of Physics
University of Oslo
Oslo, Norway

ISBN 978-3-031-77052-4   ISBN 978-3-031-77053-1 (eBook)
https://doi.org/10.1007/978-3-031-77053-1

© The Editor(s) (if applicable) and The Author(s), under exclusive license to Springer Nature Switzerland AG 2025, corrected publication 2025

This work is subject to copyright. All rights are solely and exclusively licensed by the Publisher, whether the whole or part of the material is concerned, specifically the rights of reprinting, reuse of illustrations, recitation, broadcasting, reproduction on microfilms or in any other physical way, and transmission or information storage and retrieval, electronic adaptation, computer software, or by similar or dissimilar methodology now known or hereafter developed.
The use of general descriptive names, registered names, trademarks, service marks, etc. in this publication does not imply, even in the absence of a specific statement, that such names are exempt from the relevant protective laws and regulations and therefore free for general use.
The publisher, the authors and the editors are safe to assume that the advice and information in this book are believed to be true and accurate at the date of publication. Neither the publisher nor the authors or the editors give a warranty, expressed or implied, with respect to the material contained herein or for any errors or omissions that may have been made. The publisher remains neutral with regard to jurisdictional claims in published maps and institutional affiliations.

This Springer imprint is published by the registered company Springer Nature Switzerland AG
The registered company address is: Gewerbestrasse 11, 6330 Cham, Switzerland

If disposing of this product, please recycle the paper.

Preface

This textbook is based on lecture notes from a one-semester course designed for undergraduate students in the physical sciences. The course is offered by the Department of Physics at the University of Oslo, mainly to third-year physics students. However, it has also been popular among students from astrophysics and Earth sciences.

The book is written for students who already have a basic understanding of calculus and linear algebra. It aims to help them further develop problem-solving skills for solving analytically problems in physical sciences.

The material is divided into 26 lectures, covering five key areas of applied mathematics: Complex Analysis, Variational Calculus, Ordinary Differential Equations, Integral Transformations, and Partial Differential Equations. Each lecture introduces important concepts and theorems, with a focus on applying analytical methods to solving mathematical problems.

I would like to thank Susanne Viefers, who previously taught this course, for sharing her teaching resources. My thanks also go to the teaching assistants Larissa Bravina, Vidar Skogvoll, Jonas Eidesen, David Richard Shope, and Isak Cecil Onsager Rukan for their valuable feedback in refining the course material. Finally, I would like to thank the students at the University of Oslo who have taken this course over the past three years. Their remarks, corrections, and questions have been important in the development of these lecture notes.

Oslo, Norway   Luiza Angheluta

Contents

Complex Analysis 1
1 Lecture 1: Complex Numbers 1
  1.1 Representations of Complex Numbers 2
  1.2 Application to Classical Mechanics 6
2 Lecture 2: Complex Series and Functions 6
  2.1 Complex Series 6
  2.2 Elementary Functions of One Complex Variable 9
3 Lecture 3: Analytic Functions 15
  3.1 Complex Differentiation 15
  3.2 Cauchy-Riemann Conditions 17
4 Lecture 4: Complex Integration 22
  4.1 Line Integrals 22
  4.2 Cauchy's Theorem 24
  4.3 Cauchy's Integral Formula 25
5 Lecture 5: Generalized Cauchy's Integral Formula 30
  5.1 Generalized Cauchy's Integral Theorem 30
  5.2 Cauchy's Inequality 32
6 Lecture 6: Taylor Expansion 35
  6.1 Taylor Expansion 35
7 Lecture 7: Laurent Expansion 41
  7.1 Laurent Expansion 41
  7.2 Definitions: Poles and Residue 44
  7.3 Examples 45
8 Lecture 8: Methods of Finding the Residue 51
  8.1 Finding the Residue 51
  8.2 Residue Theorem 57
9 Lecture 9: Definite Integrals 61
  9.1 Mapping to Unit Circle 61
  9.2 Extend to Complex Plane 63
  9.3 Jordan's Lemma 65
  9.4 Improper Integrals 67

Variational Calculus 73
1　Lecture 10: Calculus of Variations: Euler Stationarity Condition 73
　　1.1　Euler Condition 74
2　Lecture 11: Cycloids and Geodesics 80
　　2.1　Curve Representation $x = x(y)$ 81
　　2.2　Polar Representation $\theta = \theta(r), r = r(\theta)$ 84
　　2.3　Geodesics on Quadratic Surfaces 84
3　Lecture 12: Fermat's Principle and Hamilton's Principle 88
　　3.1　Fermat's Principle 89
　　3.2　Hamilton's Principle 91

Ordinary Differential Equations 97
1　Lecture 13: First Order ODEs 97
　　1.1　First Order ODE's 98
　　1.2　Linear Second Order Ode's 102
2　Lecture 14: Linear Second Order ODEs 105
　　2.1　Second Order Ode's with Constant Coefficients 105
3　Lecture 15: Green Function Method 115
　　3.1　Dirac Delta Function 115
　　3.2　Linear Second Order Ode's: Green Function Method 118
4　Lecture 16: Green Function Method Examples 122
　　4.1　Green Function Method 123
　　4.2　Initial Value Problem 125
　　4.3　Boundary Value Problem 127
5　Lecture 17: Power Series Method 129
　　5.1　Power Series Method 129
6　Lecture 18: Frobenius Method 138

Fourier Series 147
1　Lecture 19: Fourier Series for 2π-Periodicity 147
　　1.1　Functions with 2π Periodicity 148
2　Lecture 20: Fourier Series for $2L$-Periodicity 157
　　2.1　Even and Odd Extensions 161
　　2.2　Completeness Relation 165

Integral Transforms 169
1　Lecture 21: Fourier Transform 169
　　1.1　Fourier Transform 169
2　Lecture 22: Laplace Transform 179
　　2.1　Laplace Transform 179
　　2.2　Inverse Laplace Transform Integral (Supplementary) 187
3　Lecture 23: Convolution of Integral Transforms 188
　　3.1　Integral Transform of a Convolution 188
　　3.2　Boundary Value Problem 193
　　3.3　Initial Value Problem 195

Partial Differential Equations 199
1 Lecture 24: Separation of Variable Method 199
   1.1 Definitions 199
   1.2 Separation of Variable Method 200
   1.3 Diffusion Equation in 2D: Cartesian Coordinates 206
2 Lecture 25: Separation of Variable Method Non-Cartesian
  Coordinates 209
   2.1 Laplace Equation: Spherical Coordinates 209
   2.2 Diffusion Equation in Polar Coordinates 215
3 Lecture 26: Integral Transform Method 217
   3.1 Laplace Transform Method 218
   3.2 Fourier Transform Method 221
   3.3 Green's Function Method 224

Correction to: Analytical Methods in Physics C1

References 229

List of Figures

Complex Analysis

Fig. 1	Representation of $z = 2 + 3i$ in the complex plane	2
Fig. 2	Complex plot of the magnitude and argument of z. The colormap shows the argument of z, $\theta = arg(z)$ which has a 2π jump across the negative real axis. The shaded color represents in linear increase of the magnitude r	4
Fig. 3	Complex exponential e^z in the z-plane. The colormap shows the argument of e^z, which has periodic jumps of 2π across lines parallel to the x-axis. The shaded color represents the magnitude $|e^z| = e^x$	10
Fig. 4	Power function z^2 in the z-plane. The colormap shows the argument of z^2. The shaded color corresponds to the magnitude $|z^2| = r^2$	11
Fig. 5	Complex logarithm $\text{Ln}(z - z_0)$ in the z-plane. The colormap shows the argument, which has a branch point at z_0 (red dot). The shaded color represents the magnitude of the logarithm	13
Fig. 6	The circle $|z - 2(1 + i)| = 2$ from Example 29	23
Fig. 7	Contour isolating the point z_0	26
Fig. 8	Contour used in Exercise 30. The singular points are marked by the red dots	27
Fig. 9	Rectangular contour from Exercise 33. The singular points are marked by the red dots	30
Fig. 10	Contour $|z| = 3$ for Exercise 34	31
Fig. 11	Disk of convergence for Taylor expansion at z_0	37
Fig. 12	Disk of convergence for Taylor expansion of $\text{Ln}(z)$ around $z = 1$	39
Fig. 13	Contour circumventing z_0 and enclosing the region where $f(z)$ is analytic	42
Fig. 14	Expansion domain for Example 41	46
Fig. 15	Expansion domain for Example 42	47

Fig. 16	Expansion domain for Example 43	48
Fig. 17	Expansion domain for Example 44	49
Fig. 18	Analytic domain around isolated singularities	59
Fig. 19	Analytic domain for Example 55	60
Fig. 20	Analytic domain for Example 56	61
Fig. 21	Illustration for Example 57	63
Fig. 22	Illustration for Example 58	64
Fig. 23	Contour circumventing simple poles on the real axis	67
Fig. 24	Contour circumventing the singular point at $z_0 = -1$ from Example 60	69
Fig. 25	Key-hole contour for Example 62	72

Variational Calculus

Fig. 1	The equilibrium profile $y(x)$ of the soap film	80
Fig. 2	Surface of revolution obtain by rotating the profile $y(x)$ about the x-axis	80
Fig. 3	Cycloid curve	81
Fig. 4	The x and y coordinates of a point on the circle rolling to towards the right	83
Fig. 5	Example of a helix on the cylinder for $a = 1$ and $b = 0$	86
Fig. 6	Example of a cone geodesic for $\theta_0 = 0$ and $\kappa = 1$	88
Fig. 7	Snell's law	90

Ordinary Differential Equations

Fig. 1	Example of sequence functions approaching the Dirac delta function	116
Fig. 2	Example of Legendre polynomials	136
Fig. 3	Bessel functions $J_0(x), J_1(x), J_2(x)$ as functions of x	143

Fourier Series

Fig. 1	Periodic function given by Eq. 1	150
Fig. 2	Truncated Fourier series from Eq. 1 for $N = 5$, $N = 400$. Notice that the Fourier series overshoots around discontinuities	152
Fig. 3	Two different periodic functions starting from $f(x) = x^3$ in different basic intervals: (left) $x \in [-\pi, \pi]$, (right) $x \in [0, 2\pi]$	155
Fig. 4	Truncated Fourier series from Eq. 8	160
Fig. 5	Truncated Fourier series from Eq. (12)	160
Fig. 6	Fourier series of the even extension from Eq. (16)	162
Fig. 7	Truncated sin series from Eq. (17)	164

Partial Differential Equations

Fig. 1	Solution of the 1D wave equation for $L = 1, c = 1$ and initial conditions $f(x) = \sin(2\pi x)$ and $g(x) = 0$	204
Fig. 2	Solution of the 1D diffusion equation for $L = 1, c = 1$ and the initial profile $f(x) = \sin(2\pi x)$	206
Fig. 3	The profile of u(x,t) as function of x for different values of t	219
Fig. 4	The Gaussian profile at different times for $c = 1$ and $b = 2$	223

Complex Analysis

1 Lecture 1: Complex Numbers

Complex numbers are powerful generalizations of real numbers. Whilst real numbers are lined up on the real axis, complex numbers are defined as ordered pairs of real numbers (x, y). The order within the pair is very important, because x represents the real part of a complex number, while y is its imaginary part. The y axis takes the role as the axis of imaginary numbers. The fundamental *imaginary* number is defined by the square root of negative one

$$i = \sqrt{-1},$$

such that numbers corresponding to taking the square root of negative numbers can be represented in units of i. The imaginary number i has its conjugate number denoted by $-i$ such that their product is one,

$$i \cdot (-i) = -i^2 = 1.$$

A *complex number* is denoted as $z = x + iy$, where $x = Re(z)$ is called the real part and $y = Im(z)$ is the imaginary part. Both x and y are real numbers, and when they take these roles, we identify the (x, y) plane with the *complex plane* or the Argand diagram. The x-axis is called the *real axis* and the y-axis is called the *imaginary axis*.

Complex numbers are the constitutive elements of complex analysis which was developed as a generalisation of the real analysis. In the following chapters, we will introduce fundamental concepts and tools from complex analysis to solve integrals. First, let us get more familiar with complex numbers and elementary functions of complex numbers.

Fig. 1 Representation of $z = 2 + 3i$ in the complex plane

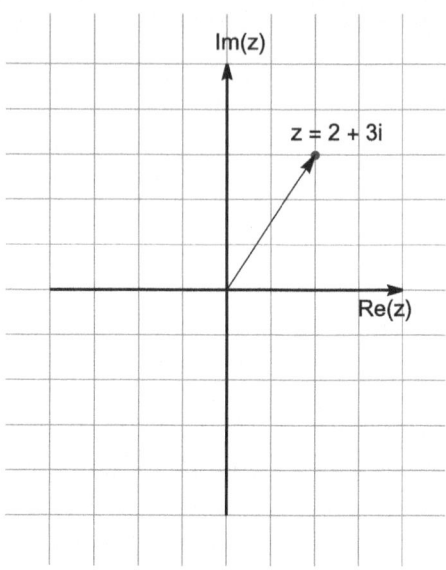

Example 1 (*Example of complex numbers in algebra*) This quadratic equation $z^2 + 4 = 0$ has the solutions given by the complex conjugate numbers $z_1 = \sqrt{-4} = 2i$, $z_2 = -\sqrt{-4} = -2i$.

Example 2 (*Example of complex number in geometry*) The complex number $z = 2 + 3i$ is represented by a point in the complex plane that has the component 2 on the real axis and 3 in the imaginary axis (see Fig. 1).

1.1 Representations of Complex Numbers

1.1.1 Rectangular Form

The representation $z = x + iy$ is called the *rectangular form* or the Cartesian representation.

A complex number $z = x + iy$ has a complex conjugate denoted $\bar{z} = x - iy$, which has the same real part and an imaginary part with opposite sign, such that their product is real-valued and given by

$$z \cdot \bar{z} = x^2 + y^2.$$

The real and imaginary parts follow from these transformations:

$$\boxed{x = \frac{z + \bar{z}}{2}, \quad y = \frac{z - \bar{z}}{2i}}$$

1 Lecture 1: Complex Numbers

Example 3 Represent the solution to this quadratic form $z^2 - 2z + 3 = 0$ as complex numbers.

Solution:
$$z = \frac{2 \pm \sqrt{4-12}}{2} = \frac{2 \pm \sqrt{-8}}{2} = 1 \pm \sqrt{-2} = 1 \pm i\sqrt{2}$$

Example 4 Write $z = \frac{1+i}{1-i}$ in the rectangular form.

Solution: We multiply and divide by $1+i$, so that the denominator becomes real:
$$z = \frac{(1+i)^2}{1^2 + 1^2} = \frac{1 + 2i + i^2}{2} \Rightarrow z = i$$

1.1.2 Polar Form

The coordinates x and y can also be transformed to polar coordinates,
$$x = r\cos\theta, \quad y = r\sin\theta.$$

The radius r is determined by the *magnitude* of z
$$r^2 = x^2 + y^2 = z \cdot \bar{z} \equiv |z|^2.$$

The angle θ (measured in radians) determines the **argument** of z up to an arbitrary 2π rotation. This is because any (2π) rotation returns to the same point in the plane. Thus, the argument of z is intrinsically multi-valued, and we can see this clearly when we represent a (unique) complex number z in a polar form

$$z = r\cos(\theta + 2\pi n) + ir\sin(\theta + 2\pi n), \quad n \in \mathbb{Z}$$

To eliminate this ambiguity, it is useful to define the argument of z as the *principal value* of θ which is in the basic interval $[-\pi, \pi)$. This is the default convention unless stated otherwise (Fig. 2). Using the basic interval, the polar form of z reduces to

$$z = r\left[\cos(\theta) + i\sin(\theta)\right], \quad \theta \in [-\pi, \pi)$$

What is θ? The complex plane is divided into four domains (quadrants) labeled counterclockwise. Depending on which quadrant the complex number is located, the principal value of the phase $\theta \in [-\pi, \pi)$ is determined as follows:

Fig. 2 Complex plot of the magnitude and argument of z. The colormap shows the argument of z, $\theta = arg(z)$ which has a 2π jump across the negative real axis. The shaded color represents in linear increase of the magnitude r

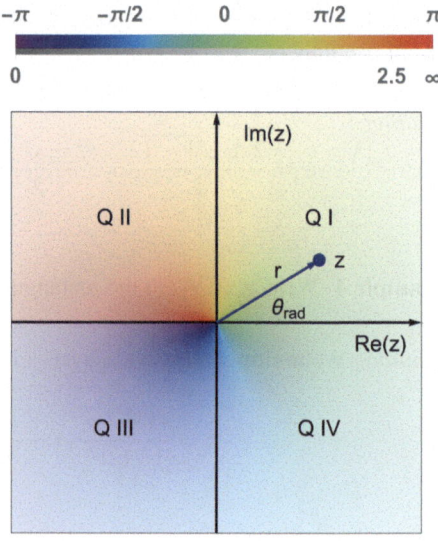

$$\theta = \begin{cases} \arctan(y/x), & x > 0, y > 0 \text{ Quadrant I} \\ \pi + \arctan(y/x), & x < 0, y > 0 \text{ Quadrant II} \\ -\pi + \arctan(y/x), & x < 0, y < 0 \text{ Quadrant III} \\ \arctan(y/x), & x > 0, y < 0 \text{ Quadrant IV} \end{cases} \quad (1)$$

1.1.3 Euler's Formula

This is a very useful formula that links trigonometric functions with the exponential for complex numbers, namely that

$$\boxed{z = r(\cos\theta + i\sin\theta) = re^{i\theta}} \quad (2)$$

The complex exponential with purely imaginary exponent has unit magnitude, namely

$$|e^{i\theta}|^2 = |\cos\theta + i\sin\theta|^2 = \cos^2\theta + \sin^2\theta = 1.$$

It also has the remarkable property that

$$e^{i\pi} = -1$$

which corresponds to a rotation by $(2k+1)\pi$ for an integer k. In other words, the number -1 represents an infinite set of equivalent complex numbers, each differing in phase by an integer multiple of 2π.

1 Lecture 1: Complex Numbers

Example 5 Write $z = -\sqrt{3} + i$ in polar form and in terms of the complex exponential.

Solution: This number is in Quadrant II because its imaginary part is positive while its real part is negative. Thus, its argument is

$$\theta = \pi + \arctan\left(-\frac{1}{\sqrt{3}}\right) = \pi - \frac{\pi}{6} = \frac{5\pi}{6}.$$

The absolute value is $r = \sqrt{3+1} = 2$. Hence, the polar form of z reads as

$$z = 2\left[\cos\left(\frac{5\pi}{6}\right) + i\sin\left(\frac{5\pi}{6}\right)\right] = 2e^{i5\pi/6}.$$

Example 6 We represent curves and domains in the complex plane by equations and inequalities. What is the curve made by the points in the complex plane satisfying $|z - 2 + i| = 2$? What is the domain of points satisfying $|z - 2 + i| \leq 2$?

Solution: The points satisfying the equality trace the circle centered at $z_0 = 2 - i$ and of radius 2. The domain satisfying this inequality is the disk enclosed by this circle.

Example 7 Represent $z = e^{i\pi} + e^{-i\pi}$ in the rectangular form.

Solution:

$$z = \cos(\pi) + i\sin(\pi) + \cos(-\pi) + i\sin(-\pi) = -2$$

Example 8 Represent $z = \frac{(i-\sqrt{3})^3}{1-i}$ in the corresponding rectangular and polar forms.

Solution:

$$z = \frac{(i - \sqrt{3})^3}{1 - i} = \frac{1+i}{2}(i^3 + (-\sqrt{3})^3 + 3i^2(-\sqrt{3}) + 3i(-\sqrt{3})^2)$$

$$= \frac{1+i}{2}(-i - 3\sqrt{3} + 3\sqrt{3} + 9i) = 4i(1+i) = -4 + 4i.$$

The modulus of z is $r = \sqrt{4^2 + 4^2} = 4\sqrt{2}$ and its argument is $\theta = \pi + \arctan(4/(-4)) = 3\pi/4$ (in QII). Thus,

$$z = 4\sqrt{2}\left[\cos\left(\frac{3\pi}{2}\right) + i\sin\left(\frac{3\pi}{2}\right)\right] = 4\sqrt{2}e^{3i\pi/4}.$$

1.2 Application to Classical Mechanics

1.2.1 Particle Motion in the Plane

The trajectory of a particle moving in the (x, y) plane can be readily expressed as a time-dependent complex number $z(t) = x(t) + iy(t)$. This is a path in the complex plane. We can compute the magnitude of velocity and acceleration in terms of the time derivatives of $z(t)$.

Example 9 Let us consider the trajectory

$$z(t) = 1 + 3e^{2it} = 1 + 3\cos(2t) + 3i\sin(2t)$$

which implies that $x(t) = 1 + 3\cos(2t)$ and $y(t) = 3\sin(2t)$.

The velocity in the complex form is determined by the time derivative of z:

$$\frac{dz}{dt} = \frac{dx}{dt} + i\frac{dy}{dt} = v_x + iv_y \Rightarrow$$
$$\frac{dz}{dt} = -6\sin(2t) + 6i\cos(2t) = 6i\left[\cos(2t) + i\sin(2t)\right] = 6ie^{2it}.$$

Therefore, the speed is determined by the magnitude of dz/dt:

$$v_x^2 + v_y^2 = \left|\frac{dz}{dt}\right| = \left|6ie^{2it}\right| = 6$$

which is independent of time. Thus, the particle is moving at a constant speed, but it changes its direction. The distance of $z(t)$ away from 1 is $|z - 1| = |3e^{2it}| = 3$ and also a constant. So the particle moves uniformly in a circle centred at point $(1, 0)$. Notice that the magnitude of acceleration is nonzero, $\left|\frac{d^2z}{d^2t}\right| = 12$ and corresponds to a non-zero centripetal acceleration.

2 Lecture 2: Complex Series and Functions

2.1 Complex Series

A complex series is defined as the sum of complex numbers

$$s_N = x_N + iy_N = \sum_{n=0}^{N}(a_n + ib_n),$$

where a_n and b_n are real numbers for every term in the series, i.e. $n = 0, 1, 2 \ldots$.

2 Lecture 2: Complex Series and Functions

For finite N, the sum s_N is a well-defined complex number. The question is if s_N converges to a well-defined sum $s = x + iy$ in the limit of $N \to \infty$, i.e.

$$s = \lim_{N \to \infty} s_N$$

For this, we use the convergence criterion of the real series formed by the real and imaginary parts based on the ratio test. The complex series is convergent when the ratio between successive coefficients satisfies that

$$\boxed{\rho = \lim_{n \to \infty} \left| \frac{a_{n+1} + ib_{n+1}}{a_n + ib_n} \right| < 1} \qquad (3)$$

Example 10 Test if $\sum_n \frac{(1+3i)^n}{n!}$ is convergent.

Solution: To check this, we evaluate the magnitude of the ratio between successive coefficients,

$$\left| \frac{(1+3i)^{n+1} n!}{(n+1)!(1+3i)^n} \right| = \left| \frac{(1+3i)}{(n+1)} \right| = \frac{|1+3i|}{n+1} = \frac{\sqrt{10}}{n+1} \to_{n \to \infty} 0$$

Thus, the series is convergent.

A **complex power series** is defined as a series of powers of z,

$$\boxed{f_N(z) = \sum_{n=0}^{N} a_n z^n = a_0 + a_1 z + a_2 z^2 + \cdots ,}$$

where a_n are complex numbers. Notice that the power series is a function of z. The domain of z for which $f_N(z)$ converges to a sum is determined by the convergence criterion:

$$\boxed{\rho = \lim_{n \to \infty} \left| \frac{a_{n+1} z}{a_n} \right| < 1} \qquad (4)$$

which corresponds to the disk of convergence in the complex plane.

Power series are very useful in real analysis where functions can be conveniently Taylor expanded as power series. Power series are perhaps even more useful in complex analysis as they map out whole regions in the complex plane where functions are well-behaved.

Example 11 Find the radius of convergence for this power series

$$f_N(z) = \sum_{n=0}^{N} \frac{(iz)^n}{n^2}$$

Solution: The coefficient of z^n is i^n/n^2. Let us look at the convergence criterion:

$$\lim_{n\to\infty} \left|\frac{izn^2}{(n+1)^2}\right| = |iz| \lim_{n\to\infty} \frac{n^2}{(n+1)^2} = |iz| < 1$$

Thus, the convergence condition is $|iz| = |z| < 1$, corresponding to the unit disk centered at the origin.

Example 12 Find the radius of convergence of the following power series:

$$f_N(z) = \sum_{n=0}^{N} \frac{z^n}{n!}$$

Solution: Let us look at the convergence criterion:

$$\lim_{n\to\infty} \left|\frac{z}{(n+1)}\right| = |z| \lim_{n\to\infty} \frac{1}{(n+1)} \to 0$$

This power series is convergent for any value of z. Based on the analogy with real power series, we may recognise this as the power series expansion of the complex exponential.

Example 13 Prove Euler's formula

$$\cos\theta + i\sin\theta = e^{i\theta}$$

Solution: Let us use the Taylor expansion of the trigonometric real functions:

$$\cos\theta = 1 - \frac{\theta^2}{2!} + \frac{\theta^4}{4!} \cdots$$

$$\sin\theta = \theta - \frac{\theta^3}{3!} + \frac{\theta^5}{5!} \cdots$$

Combing them in the complex form and re-arranging the terms:

$$\cos\theta + i\sin\theta = \left(1 - \frac{\theta^2}{2!} + \frac{\theta^4}{4!}\right) + i\left(\theta - \frac{\theta^3}{3!} + \frac{\theta^5}{5!} \cdots\right)$$

$$= 1 + i\theta + \frac{(i\theta)^2}{2!} + \frac{(i\theta)^3}{3!} \cdots$$

$$= e^{i\theta}$$

2 Lecture 2: Complex Series and Functions

Example 14 Find the disk of convergence for

$$\sum_{n=0}^{\infty} \left(\frac{z-2+i}{2}\right)^n$$

Solution: The radius of convergence is

$$\rho = \lim_{n \to \infty} \left| \frac{(z-2+i)^{n+1}/2^{n+1}}{(z-2+i)^n/2^n} \right| = \frac{|z-2+i|}{2} < 1$$

This power series is convergent when the complex number z satisfies the inequality $|z - 2 + i| < 2$. Writing this in the rectangular form, we see that it corresponds to $|(x - 2) + i(y + 1)| < 2$, which is the disk centered at the point $(2, -1)$ and of radius 2.

2.2 Elementary Functions of One Complex Variable

A complex function $f(z)$ is a map that takes one complex number $z = x + iy$ and returns another complex number. The rectangular form of $f(z)$ can be written as

$$\boxed{f(z) = u(x, y) + iv(x, y)}$$

where $u(x, y) = Re(f(z))$ is the real part and $v(x, y) = Im(f(z))$ is the imaginary part. This is also called the *normal form*. Just like with complex numbers, the functions u and v need to be real-valued functions!

Definition 1 $f(z)$ is single-valued when $f(z)$ takes a unique value for each z. In polar form, this implies restricting the argument of z to the basic interval, i.e. $\theta \in [-\pi, \pi)$.

Next, we introduce important elementary functions and their corresponding normal form (Fig. 3).

2.2.1 Exponential Function

The exponential holds a special place in complex analysis through the powerful Euler's formula connecting the exponential with purely imaginary exponent with trigonometric functions

$$e^{i\theta} = \cos(\theta) + i\sin(\theta).$$

Fig. 3 Complex exponential e^z in the z-plane. The colormap shows the argument of e^z, which has periodic jumps of 2π across lines parallel to the x-axis. The shaded color represents the magnitude $|e^z| = e^x$

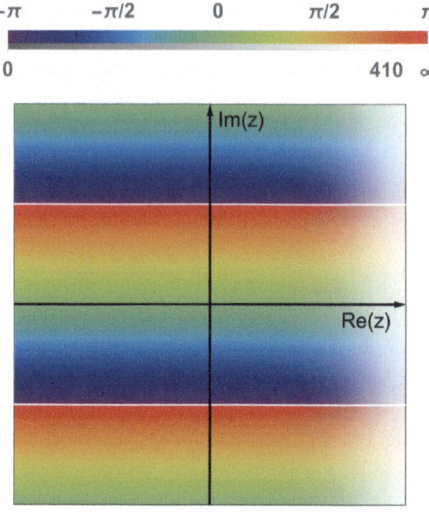

Using the rectangular form $z = x + iy$, we can also write the exponential of z as

$$e^z = e^x e^{iy} = e^x(\cos y + i \sin y), \tag{5}$$

where e^x is the familiar, non-negative defined exponential that blows up for positive values of x or decreases to zero for negative x. However, the complex exponential has a more subtle behavior in the plane due to the trigonometric dependence on y.

Another common complex exponential is e^{iz} which can be expressed using Euler's formula as

$$e^{iz} = \cos(z) + i \sin(z)$$

and relates to the complex trigonometric functions as discussed further below (Fig. 4).

Some useful identities:

$$\boxed{e^z e^{\bar{z}} = e^{2x}} \qquad \boxed{e^z e^{-\bar{z}} = e^{2iy}}$$

Example 15 Write $z = e^{-i\pi/4 + \ln 3}$ in the rectangular form.

Solution: $z = e^{-i\pi/4 + \ln 3} = 3\left(\cos\frac{\pi}{4} - i\sin\frac{\pi}{4}\right) = \frac{3}{\sqrt{2}} - i\frac{3}{\sqrt{2}}$.

2.2.2 Power and Root Functions

The power function takes a complex number to a power n:

$$f(z) = z^n, \quad n \in \mathbb{Z}$$

Fig. 4 Power function z^2 in the z-plane. The colormap shows the argument of z^2. The shaded color corresponds to the magnitude $|z^2| = r^2$

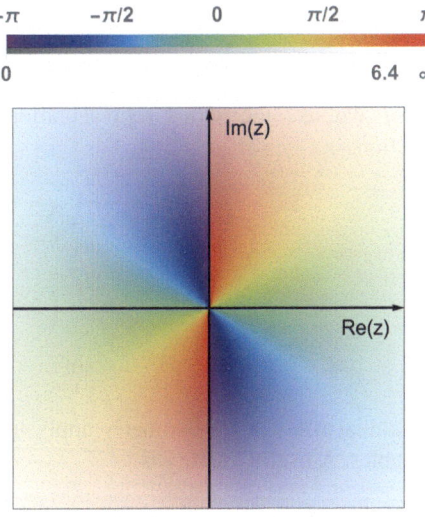

De Moivre Formula: This is a useful identity that follows from the Euler's formula and provides a trigonometric (polar) form of the power function:

$$\boxed{z^n = r^n[\cos(n\theta) + i\sin(n\theta)]}$$

This relation follows straightforwardly from the Euler's formula, namely

$$z^n = (re^{i\theta})^n = r^n e^{in\theta} = r^n[\cos(n\theta) + i\sin(n\theta)].$$

Similarly, the root function takes the n-th root of a complex number:

$$f(z) = z^{1/n}, \quad n \in \mathbb{Z}, \quad n \neq 0$$

It has the corresponding polar form through De Moivre formula as

$$\boxed{z^{1/n} = r^{1/n}[\cos(\theta/n) + i\sin(\theta/n)]}$$

2.2.3 Trigonometric Functions

We have useful representations of $\sin z$ and $\cos z$ in terms of the complex exponential function:

$$\boxed{\sin z = \frac{e^{iz} - e^{-iz}}{2i}} \tag{6}$$

$$\cos z = \frac{e^{iz} + e^{-iz}}{2} \qquad (7)$$

Proof We use the series representation of the exponential and trigonometric functions

$$\frac{e^{iz} - e^{-iz}}{2i} = \frac{1}{2i} \sum_n \frac{(iz)^n(1-(-1)^n)}{n!}$$

$$= \frac{1}{2i}\left(2iz - 2i\frac{z^3}{3!} + 2i\frac{z^5}{5!}\cdots\right)$$

$$= \sin z$$

Similar rules of trigonometry apply to the trigonometric functions of complex variables. A useful identity is:

$$\boxed{\sin^2 z + \cos^2 z = 1}$$

Proof

$$\sin^2 z = \left(\frac{e^{iz} - e^{-iz}}{2i}\right)^2 = \frac{e^{2iz} + e^{-2iz} - 2}{-4}$$

$$\cos^2 z = \left(\frac{e^{iz} + e^{-iz}}{2}\right)^2 = \frac{e^{2iz} + e^{-2iz} + 2}{4}$$

By adding them up, we get that

$$\sin^2 z + \cos^2 z = 1$$

2.2.4 Hyperbolic Functions

Like with the trigonometric functions, we have useful representations of the hyperbolic functions in terms of the complex exponentials:

$$\boxed{\sinh z = \frac{e^z - e^{-z}}{2}} \qquad (8)$$

$$\boxed{\cosh z = \frac{e^z + e^{-z}}{2}} \qquad (9)$$

2 Lecture 2: Complex Series and Functions

Useful Identity 1:
$$\cosh^2 z - \sinh^2 z = 1 \tag{10}$$

Proof
$$\sinh^2 z = \left(\frac{e^z - e^{-z}}{2}\right)^2 = \frac{e^{2z} + e^{-2z} - 2}{4}$$

$$\cosh^2 z = \left(\frac{e^z + e^{-z}}{2}\right)^2 = \frac{e^{2z} + e^{-2z} + 2}{4}$$

By subtracting them, we have that
$$\cosh^2 z - \sinh^2 z = 1$$

Useful Identity 2:
$$\boxed{\sinh iz = i \sin z} \tag{11}$$

Proof We have that
$$\sinh iz = \frac{e^{iz} - e^{-iz}}{2} = i \sin z$$

2.2.5 Logarithmic Function

The natural representation of the logarithm is using the polar form z namely (Fig. 5)

$$\boxed{\ln(z) = \ln r + i(\theta + 2\pi n).}$$

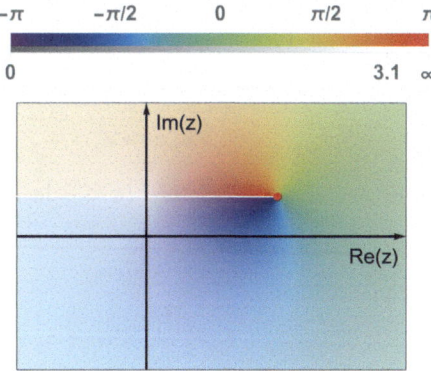

Fig. 5 Complex logarithm $\text{Ln}(z - z_0)$ in the z-plane. The colormap shows the argument, which has a branch point at z_0 (red dot). The shaded color represents the magnitude of the logarithm

Notice that it is multi-valued through its linear dependence on the argument of z. The single-valued logarithm is rendered by constraining θ to its principal value interval:

$$\boxed{\operatorname{Ln}(z) = \ln(r) + i\theta, \qquad \theta \in (-\pi, \pi].}$$

Example 16 Write $\operatorname{Ln}(1 + i\sqrt{3})$ in the rectangular form.

Solution: The modulus is $r = |z| = 2$ and the argument is $\theta = \arctan(\sqrt{3}) = \pi/3$. Hence,

$$\operatorname{Ln}(z) = \ln(2) + i\frac{\pi}{3}.$$

2.2.6 Complex Roots and Powers

The exponent of root and power functions can be a complex number,

$$\boxed{z^w = e^{w \ln z}}$$

where both z and w are complex number.

Example 17 Write $z = (1 + i)^{1-i}$ in the polar form.

Solution: Let us write this in terms of the complex exponential as

$$z = e^{(1-i)\operatorname{Ln}(1+i)}$$

where

$$\operatorname{Ln}(1 + i) = \ln\sqrt{2} + i(\pi/4).$$

Thus,

$$(1 - i)\operatorname{Ln}(1 + i) = (1 - i)[\ln\sqrt{2} + i(\pi/4)] = \ln\sqrt{2} + \pi/4 + i(-\ln\sqrt{2} + \pi/4)$$

and

$$\begin{aligned} z &= \sqrt{2} e^{\pi/4} e^{i(-\ln\sqrt{2} + \pi/4)} \\ &= \sqrt{2} e^{\pi/4} [\cos(\pi/4 - \ln\sqrt{2}) + i\sin(\pi/4 - \ln\sqrt{2})]. \end{aligned}$$

3 Lecture 3: Analytic Functions

Now that we are more familiar with complex numbers and functions, we are ready to embark on complex differentiation and learn about basic properties of analytic functions.

3.1 Complex Differentiation

Definition 2 (*Complex derivative*) The complex derivative of $f(z)$ at a point z is defined as
$$f'(z_0) = \lim_{z \to z_0} \frac{f(z) - f(z_0)}{z - z_0}. \tag{12}$$

Unlike on the real line, there are infinitely many possibilities to construct increments $\Delta z = z - z_0$ in the complex plane. This changes profoundly the meaning of a derivative and its consequences on the differentiability of $f(z)$.

The complex function $f(z)$ has a derivative at a point z_0 if $f'(z_0)$ has a unique value, i.e. $f'(z_0)$ is independent on the path of vanishing Δz.

Complex differentiation has similar rules as in real analysis.

Sum Rule: $\frac{d}{dz}(f + g) = f'(z) + g'(z)$

Product Rule: $\frac{d}{dz}(f \cdot g) = f'(z)g(z) + f(z)g'(z)$

Chain Rule: $\frac{d}{dz} f(g(z)) = f'(g)g'(z)$.

Definition 3 (*Analytic function*) The complex function $f(z)$ is analytic in a domain, if it is differentiable at every point in that domain. The property of being analytic requires that the function is <u>differentiable at every point in a given domain</u>.

This is a fundamental property with important implications in complex analysis. One such implication, is that the real and imaginary parts of $f(z)$ are related with each other through the <u>Cauchy-Riemann conditions</u>. Also, analytic functions have well-defined derivatives of any order.

Example 18 Show that $f(z) = e^z$ is analytic in the entire complex plane.

Solution: The derivative of the complex exponential at an arbitrary point follows as

$$\begin{aligned}\frac{d}{dz}e^z &= \lim_{\Delta z \to 0} \frac{e^{z+\Delta z} - e^z}{\Delta z} \\ &= e^z \lim_{\Delta z \to 0} \frac{e^{\Delta x}(\cos \Delta y + i \sin \Delta y) - 1}{\Delta x + i \Delta y}.\end{aligned}$$

Since Δx and Δy are infinitesimal real-valued increments, we can Taylor expand to first order the trigonometric functions and the real exponential,

$$\frac{d}{dz}e^z = e^z \lim_{\Delta z \to 0} \frac{(1+\Delta x)(1+i\Delta y) - 1}{\Delta x + i\Delta y}$$
$$= e^z \lim_{\Delta z \to 0} \frac{\Delta x + i\Delta y}{\Delta x + i\Delta y}$$
$$= e^z$$

This calculation is independent of the value of z, hence the complex exponential is differential in the complex plane hence analytic.

Example 19 Show that $\bar{z}$ is not differentiable.

Solution: We evaluate the derivative at an arbitrary point z as

$$\frac{d}{dz}\bar{z} = \lim_{\Delta z \to 0} \frac{(\bar{z} + \Delta \bar{z}) - \bar{z}}{\Delta z}$$
$$= \lim_{\Delta z \to 0} \frac{\Delta x - i\Delta y}{\Delta x + \Delta y}$$

For any z, if we approach it long the x-axis ($\Delta y = 0$), then the derivative is 1. However, if we approach it along the y-axis ($\Delta x = 0$), then the same derivative takes the value -1. The derivative takes different values depending on which direction we approach z. Hence, $\bar{z}$ is not differentiable.

Example 20 Show that $|z|^2$ is not analytic.

Solution:

$$\frac{d}{dz}|z|^2 = \lim_{\Delta z \to 0} \frac{|z + \Delta z|^2 - |z|^2}{\Delta z}$$
$$= \lim_{\Delta z \to 0} \frac{(z + \Delta z)(\bar{z} + \Delta \bar{z}) - |z|^2}{\Delta z}$$
$$= \lim_{\Delta z \to 0} \frac{z\Delta \bar{z} + \bar{z}\Delta z}{\Delta z}$$
$$= \lim_{\Delta z \to 0} \frac{2x\Delta x + 2y\Delta y}{\Delta x + i\Delta y}$$

The numerator is real, while the denominator is a complex number. This is a problem! Let us approach the point z along the x-axis (i.e. $\Delta y = 0$), then the derivative on this path equals $2x$. On the other hand, if we choose the path along the y-axis (i.e. $\Delta x = 0$), then the derivative is instead equal to $-2iy$. This tells us that the value of $f'(z)$ is path-dependent. However, at $z_0 = 0$, we see that the derivative vanishes

3 Lecture 3: Analytic Functions

irrespective of the path. Thus, $|z|^2$ has a complex derivative only at $z = 0$, hence it is not analytic.

Definition 4 (*Regular point*) A point z_0 is said to be a regular point when $f(z)$ is analytic at z_0, i.e. we can determine a region around z_0 (neighborhood) where $f(z)$ is differentiable at every point.

Definition 5 (*Singular point*) A point of z_0 is singular when $f(z)$ is not analytic at z_0, i.e. there is no neighborhood of z_0 where $f(z)$ is differentiable.

Definition 6 (*Isolated singularity*) A point z_0 is an isolated singularity when $f(z)$ is analytic everywhere else except inside a small disk centered at z_0, i.e. for $|z - z_0| \leq \epsilon$.

3.2 Cauchy-Riemann Conditions

Theorem 1 *If a complex function $f(z) = u(x, y) + iv(x, y)$ is analytic in a given domain, then $u(x, y)$ and $v(x, y)$ satisfy the **Cauchy-Riemann conditions** in that region,*

$$\boxed{\frac{\partial u}{\partial x} = \frac{\partial v}{\partial y}, \quad \frac{\partial v}{\partial x} = -\frac{\partial u}{\partial y}} \tag{13}$$

Proof Let us start from the definition of the derivative

$$f'(z) = \lim_{\Delta z \to 0} \frac{f(z + \Delta z) - f(z)}{\Delta z} = \frac{du + idv}{dx + idy}.$$

For $dx = 0$, the derivative becomes

$$f'(z) = \lim_{dy \to 0} \frac{du + idv}{idy} = -i\frac{\partial u}{\partial y} + \frac{\partial v}{\partial y}.$$

Similarly if we now take $dy] = 0$, the derivative equals to

$$f'(z) = \lim_{dx \to 0} \frac{du + idv}{dx} = \frac{\partial u}{\partial x} + i\frac{\partial v}{\partial x}.$$

From the uniqueness of $f'(z)$, the two expressions must be equal and this leads the Cauchy-Riemann conditions.

Alternatively, we can use the rule of differentiation of $f(z)$ as an implicit function of x and y through $z = x + iy$:

$$\partial_x f = \frac{\partial f(z)}{\partial x} = \frac{df}{dz}\frac{\partial z}{\partial x} = f'(z), \quad \partial_y f = \frac{\partial f(z)}{\partial y} = \frac{df}{dz}\frac{\partial z}{\partial y} = if'(z)$$

Since $f(z)$ is analytic, it follows that $f'(z)$ is uniquely defined, hence the two expressions lead to
$$\partial_x f = -i\partial_y f$$

Evaluating the partial derivatives of $f(z)$ in terms of the partial derivatives of $u(x, y)$ and $v(x, y)$, we find equivalent expressions of $f'(z)$:

$$\boxed{f'(z) = \partial_x u + i\partial_x v = -i\partial_y u + \partial_y v} \tag{14}$$

Taking the real and imaginary parts, we arrive at the Cauchy-Riemann relations.

Theorem 2 *If $u(x, y)$ and $v(x, y)$ and their partial derivatives with respect to x and y are differentiable and satisfy the Cauchy-Riemann (C-R) conditions in a region of the complex plane, then $f(z)$ is analytic inside that region (not necessarily on the boundary).*

Proof We want to show that $f'(z)$ is uniquely defined (path-independent) for each point in the region where $u(x, y)$ and $v(x, y)$ and their derivatives are continuous and C-R conditions are satisfied. Since $u(x, y)$ and $v(x, y)$ are differentiable, their infinitesimal forms are

$$du = \partial_x u\, dx + \partial_y u\, dy$$
$$dv = \partial_x v\, dx + \partial_y v\, dy.$$

Inserting this back into the differentiation formula and rearranging terms, we get

$$\frac{df}{dz} = \frac{(\partial_x u + i\partial_x v)dx + (\partial_y u + i\partial_y v)dy}{dx + i\,dy}$$

By using the C-R relations, it follows that

$$\frac{df}{dz} = \frac{(\partial_x u + i\partial_x v)dx + (-\partial_x v + i\partial_x u)dy}{dx + i\,dy}$$
$$= \frac{(\partial_x u + i\partial_x v)dx + i(\partial_x u + i\partial_x v)dy}{dx + i\,dy}$$
$$= \partial_x u + i\partial_x v$$

which is well-defined since u and v are differentiable functions.

Example 21 Show that $f(z) = \frac{1}{z} = \frac{\bar{z}}{|z|^2}$ is analytic everywhere outside the origin $z_0 = 0$.

Solution: First, we rewrite $f(z) = \frac{\bar{z}}{|z|^2}$ in the canonical form to determine u and v functions, namely:

3 Lecture 3: Analytic Functions

$$u(x, y) = \frac{x}{x^2 + y^2}, \quad v(x, y) = -\frac{y}{x^2 + y^2}$$

By compute the corresponding partial derivatives

$$\partial_x u(x, y) = \frac{-x^2 + y^2}{(x^2 + y^2)^2}, \quad \partial_y u(x, y) = -\frac{2xy}{(x^2 + y^2)^2}$$

$$\partial_y v(x, y) = \frac{-x^2 + y^2}{(x^2 + y^2)^2}, \quad \partial_x v(x, y) = \frac{2xy}{(x^2 + y^2)^2},$$

we notice that the partial derivatives are well-defined outside the origin $z_0 = 0$ where they satisfy the C-R conditions, i.e. $\partial_x u(x, y) = \partial_y v(x, y)$ and $\partial_y u(x, y) = -\partial_x v(x, y)$. Thus the function is differentiable outside the origin, hence it is analytic.

Example 22 Similarly you can show that the simple fraction $f(z) = \frac{1}{z-z_0}$ satisfies the C-R conditions everywhere except at the isolated singular point z_0. Hence, it is analytic outside its isolated singularity. Simple fractions are a convenient way to presented isolated singularities.

Example 23 Let $f(z) = u + iv$ be an analytic function, and let $\vec{F} = v\vec{i} + u\vec{j}$ be a vector field with the components determined by the real and imaginary parts of $f(z)$. Show that div $\vec{F} = 0$ and curl $\vec{F} = 0$ correspond to C-R conditions.

Solution: The C-R relations follow straightforwardly from the definitions of divergence and curl of a 2D vector field:

$$\text{div } \vec{F} = 0 \rightarrow \partial_x v + \partial_y u = 0$$

$$\text{curl } \vec{F} = 0 \rightarrow \partial_x u - \partial_y v = 0.$$

Example 24 Let also consider an example where we use the C-R conditions to check the differentiability of a function at a point. Show that $f(z) = |z|^2$ is differentiable only at the origin.

Solution: Since,
$$u(x, y) = x^2 + y^2, \quad v(x, y) = 0$$

it follows that
$$\partial_x u = 2x, \partial_y u = 2y$$

which vanish only $x = 0, y = 0$. Thus, the C-R conditions are valid only at this point.

The property of being analytic is manifested by the Cauchy-Riemann conditions being satisfied over a *domain*. Thus, the function needs to be differentiable over a domain, but just pointwise. This implies that u and v are harmonic functions.

Definition 7 (*Harmonic function*) A real function $\phi(x, y)$ that satisfies the Laplace equation in two dimensions, i.e.

$$\nabla^2 \phi = \partial_x^2 \phi + \partial_y^2 \phi = 0,$$

is called a **harmonic function**.

Theorem 3 *(Part 1) If $f(z) = u + iv$ is analytic in a region, then u and v are conjugate harmonic functions, i.e. they satisfy the Laplace's equation in that region.*

(Part 2) Any function u (or v) satisfying the Laplace's equation in a simply-connected region is the real (or imaginary part) of an analytic function.

Proof of Part 1: We use that C-R conditions satisfied by u and v and differentiate them once more.

$$\partial_x^2 u \equiv \frac{\partial}{\partial x}\frac{\partial u}{\partial x} = \frac{\partial}{\partial x}\frac{\partial v}{\partial y}, \qquad \partial_y^2 u \equiv \frac{\partial}{\partial y}\frac{\partial u}{\partial y} = -\frac{\partial}{\partial y}\frac{\partial v}{\partial x}$$

Hence, $\partial_x^2 u + \partial_y^2 u = 0$. Similarly,

$$\partial_x^2 v \equiv \frac{\partial}{\partial x}\frac{\partial v}{\partial x} = -\frac{\partial}{\partial x}\frac{\partial u}{\partial y}, \qquad \partial_y^2 v \equiv \frac{\partial}{\partial y}\frac{\partial v}{\partial y} = \frac{\partial}{\partial y}\frac{\partial u}{\partial x}$$

Hence $\partial_x^2 v + \partial_y^2 v = 0$.

Example 25 Show that the corresponding u and v of the analytic function $f(z) = \frac{1}{z}$ are conjugate harmonic functions.

Solution: We have shown in Example 22 that $1/z$ is an analytic function for $|z| > 0$. Here we show that

$$u(x, y) = \frac{x}{x^2 + y^2}, \qquad v(x, y) = -\frac{y}{x^2 + y^2}$$

are harmonic functions. Let's differentiate $u(x, y)$:

$$\partial_x^2 u(x, y) = \partial_x \left(\frac{-x^2 + y^2}{(x^2 + y^2)^2} \right) = \frac{2x(x^2 - 3y^2)}{(x^2 + y^2)^3},$$

$$\partial_y^2 u(x, y) = -\partial_y \left(\frac{2xy}{(x^2 + y^2)^2} \right) = -\frac{2x(x^2 - 3y^2)}{(x^2 + y^2)^3}$$

Hence $\nabla^2 u(x, y) = 0$. Similarly, you can show that $\nabla^2 v(x, y) = 0$.

3 Lecture 3: Analytic Functions

Cauchy-Riemann Conditions in Polar Form: For some $f(z)$, it may be difficult to separate $u(x, y)$ and $v(x, y)$ using the rectangular form of z. In such cases, we resort on the polar form $z = re^{i\theta}$, with r and θ as the natural coordinates of u and v

$$\boxed{f(z) = u(r, \theta) + iv(r, \theta)}$$

The corresponding Cauchy-Riemann conditions in polar form read as:

$$\boxed{\frac{\partial u}{\partial r} = \frac{1}{r}\frac{\partial v}{\partial \theta}, \quad \frac{\partial v}{\partial r} = -\frac{1}{r}\frac{\partial u}{\partial \theta}} \tag{15}$$

Let us take the implicit derivative with respect to r and θ:

$$\partial_r f(z) = f'(z)\frac{dz}{dr} = \frac{z}{r}f'(z), \quad \partial_\theta f(z) = f'(z)\frac{dz}{d\theta} = izf'(z)$$

Since $f(z)$ is analytic, the two expressions for $f'(z)$ must be identical:

$$\frac{r}{z}(\partial_r u + i\partial_r v) = -i\frac{1}{z}(\partial_\theta u + i\partial_\theta v)$$

Separating out the real and imaginary parts, we obtain the C-R relations in polar form.

Example 26 Show that $f(z) = z^n$ is analytic function.

Solution: Using de Moivre formula for the power function, we have that

$$f(z) = r^n[\cos(n\theta) + i\sin(n\theta)] \Rightarrow u(r, \theta) = r^n\cos(n\theta), \quad v(r, \theta) = r^n\sin(n\theta)$$

By taking their partial derivatives, we obtain

$$\partial_r u(r, \theta) = nr^{n-1}\cos(n\theta) = \frac{1}{r}\partial_\theta v(r, \theta)$$

$$\partial_r v(r, \theta) = nr^{n-1}\sin(n\theta) = -\frac{1}{r}\partial_\theta u(r, \theta)$$

We deduce that this power function is analytic everywhere in the complex plane for $n \geq 1$.

4 Lecture 4: Complex Integration

In this lecture, we introduce complex integration and the method of curve parameterization to evaluate complex line integrals. Then, we discuss the Cauchy's theorem and Cauchy's integral formula for contour integrals.

4.1 Line Integrals

Let Γ be a simple curve[1] in the complex plane. The integral of $f(z)$ on the path Γ is defined as

$$\boxed{I = \int_\Gamma f(z)dz} \tag{16}$$

We can solve this complex integral by specifying an appropriate parameterization of Γ, that allows us to write the complex variable z as function of the curve parameter. This is analogous to how we solve line integrals in vector analysis. We illustrate this method by examples.

Example 27 Compute

$$I = 2\int_\Gamma \frac{1}{1-4z^2}dz,$$

where Γ is the positive imaginary axis (y-axis).

Solution: We parameterize Γ as $z(y) = iy$, where $y > 0$. We have that $dz = idy$. Hence, the complex integral can be transformed into the definite integral

$$I = 2\int_0^\infty \frac{idy}{1-4(iy)^2} = i\int_0^\infty \frac{2dy}{1+(2y)^2}$$
$$= i\arctan(2y)|_0^\infty$$
$$= \frac{i\pi}{2}$$

Example 28 Compute

$$I = \int_\Gamma \bar{z}dz,$$

along the straight Γ given by $z(t) = t + it$ with $t \in [0, 1]$.

---

[1] A simple curve is one which does not cross itself.

Fig. 6 The circle
$|z - 2(1+i)| = 2$ from
Example 29

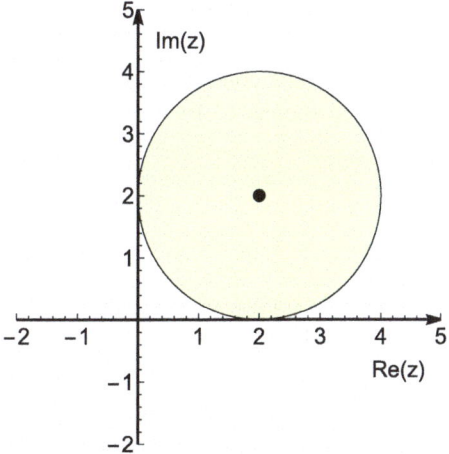

Solution: By the change of coordinate $z = (1+i)t$, with $t \in [0, 1]$, the integral becomes

$$I = (1+i)(1-i) \int_0^1 t \, dt = 1.$$

Let us now consider the contour integration over a circle (Fig. 6).

Example 29 Compute

$$I = \oint_\Gamma \frac{1}{z - 2 - 2i} dz,$$

along the circle of radius 2 and centered at $z_0 = 2 + 2i$, i.e. $\Gamma : |z - z_0| = 2$.

Solution: We parameterize the circle by the angle as $z(\theta) - z_0 = 2e^{i\theta}$, with $\theta \in [-\pi, \pi]$. Thus, $dz = 2ie^{i\theta} d\theta$ and the integral becomes

$$I = \int_{-\pi}^{\pi} \frac{2ie^{i\theta}}{2e^{i\theta}} d\theta = 2\pi i.$$

It turns out that the Cauchy's theorem is very powerful in evaluating contour integrals and circumvents the need to specifically parameterise the curve and perform the line integration. This is what we will focus on next.

4.2 Cauchy's Theorem

Definition 8 The contour integral of $f(z)$ is the line integral on a closed simple curve called contour C. By default, the contour is traversed **counter-clockwise**.

$$I = \oint_C f(z)dz \tag{17}$$

The basic example of a contour is the circle centered at a point z_0 and of radius r and it is represented as $|z - z_0| = r$. By virtue of Cauchy's theorem, the contour could have arbitrary shape.

Theorem 4 *Cauchy's theorem:* Let C be a simple contour with a continuously turning tangent except possibly at a finite number of points (corners). If $f(z)$ is **analytic** on and inside C, then the contour integral on C vanishes

$$\oint_C f(z)dz = 0 \tag{18}$$

Proof This contour integral can be written in terms of two real contour integrals as follows:

$$\oint_C (u+iv)(dx+idy) = \oint_C (udx - vdy) + i\oint_C (vdx + udy)$$

Using Green's theorem for the line integral

$$\oint_C \vec{F} \cdot d\vec{l} = \iint_D \text{curl}(\vec{F})dxdy = \iint_D (\partial_x F_y - \partial_y F_x)dxdy,$$

we can express the above real contour integrals as area integrals

$$\oint_C (udx - vdy) = \iint_D (-\partial_x v - \partial_y u)dxdy \tag{19}$$

$$\oint_C (vdx + udy) = \iint_D (\partial_x u - \partial_y v)dxdy \tag{20}$$

where D is the domain enclosed by C.

4 Lecture 4: Complex Integration

When $f(z)$ is analytic over the domain D, the Cauchy-Riemann conditions are satisfied

$$\partial_x u = \partial_y v, \qquad \partial_y u = -\partial_x v. \tag{21}$$

This implies that the integrands of the area integrals vanish and thus the integrals also vanish. Since both the real and the imaginary part of the contour integral are zero, it follows that

$$\oint_C dz f(z) = 0.$$

The Cauchy's theorem is, in many ways, the bedrock theorem on which complex analysis is build. It has many ramifications and subsequent theorems deduced from it. One important implication is that it allows us to perform integrals of complex functions with singularities inside the integration domain. The simplest case is that of an isolated singularity at z_0, which is described by the simple fraction $\frac{1}{z-z_0}$. The Cauchy's integral formula allows us to quickly evaluate contour integrals of functions that have a simple pole at z_0 and can be cast into this form as

$$\oint_C \frac{f(z)}{z - z_0} dz$$

where $f(z)$ is analytic inside the integration domain. It also provides us with a natural integral representation of complex functions.

4.3 Cauchy's Integral Formula

Theorem 5 *If $f(z)$ is **analytic** on and inside a simple contour C, then the value of $f(z)$ at a point $z = z_0$ inside the domain bounded by C is given by the following contour integral along C traversed counterclockwise:*

$$\boxed{f(z_0) = \frac{1}{2\pi i} \oint_C \frac{f(z)}{z - z_0} dz, \quad z_0 \text{ inside } C} \tag{22}$$

Since z_0 is an arbitrary point inside C, it can be treated as a complex variable. Consequently, the function $f(z)$ can be represented as a contour integral for an arbitrary z:

$$\boxed{f(z) = \frac{1}{2\pi i} \oint_C \frac{f(w)}{w - z} dw, \quad z \text{ inside } C} \tag{23}$$

Fig. 7 Contour isolating the point z_0

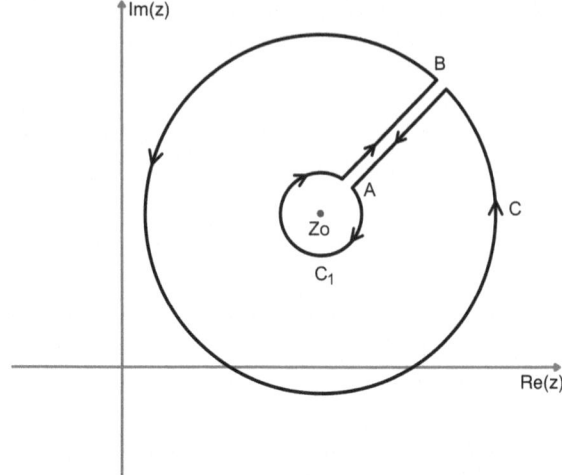

Proof Let us define the integrand as

$$\phi(z) = \frac{f(z)}{z - z_0}.$$

The function $\phi(z)$ is analytic everywhere except in a small neighborhood around z_0, making z_0 an isolated singularity. We can define a small circle C_1 that encloses z_0, allowing us to exclude z_0 from the integration domain, as shown in Fig. 7. This enables us to express the large contour C_0 as the union of C, C_1, and the line segments AB and BA. The line integrals over the segments AB and BA cancel each other out, leaving the only non-zero contribution from the contour integral around the circle C_1 as its radius approaches zero.

By Cauchy's theorem, it follows that

$$\oint_{C_0} \phi(z) dz = 0$$

which implies that

$$\oint_C \phi(z) dz + \oint_{C_1 \text{ clockwise}} \phi(z) dz = 0,$$

where we have used that the line integrals along the two incisions cancel out because they have identical integrands and the segments are traversed in opposite directions: $\int_{AB} \phi dz = -\int_{BA} \phi dz$. We can rewrite the equation above, by taking both contours counterclockwise as

$$\oint_C \phi(z) dz = \oint_{C_1} \phi(z) dz.$$

4 Lecture 4: Complex Integration

Let us represent the inner contour C_1 by a small circle centered at z_0, $z = z_0 + \epsilon e^{i\theta}$ such that an infinitesimal line element on the circle is given as $dz = \epsilon i e^{i\theta} d\theta$. Now we can perform this contour integral using the contour parameterization as

$$\oint_C \phi(z) dz = \oint_{C_1} \phi(z) dz$$
$$= \int_0^{2\pi} \phi(z_0 + \epsilon e^{i\theta}) \rho i e^{i\theta} d\theta$$
$$= \int_0^{2\pi} \frac{f(z_0 + \epsilon e^{i\theta})}{\epsilon e^{i\theta}} \epsilon i e^{i\theta} d\theta$$
$$= i \int_0^{2\pi} f(z_0 + \epsilon e^{i\theta}) d\theta$$
$$\underset{\epsilon \to 0}{=} i f(z_0) \int_0^{2\pi} d\theta$$
$$= 2\pi i f(z_0).$$

Example 30 Consider

$$I = \oint_C \frac{1}{z^2 - 1} dz,$$

where C is a counterclockwise directed contour including $z = -1$ but not $z = 1$. For simplicity, we may take the contour as the circle illustrated in Fig. 8.

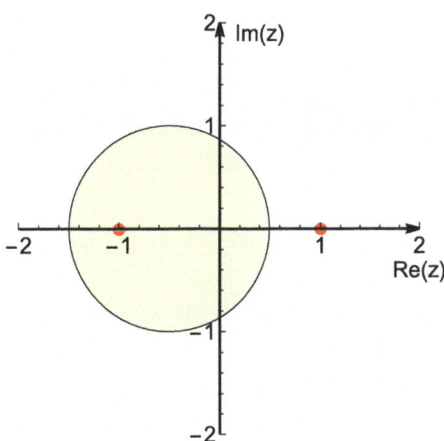

Fig. 8 Contour used in Exercise 30. The singular points are marked by the red dots

Solution: This is our first example where we use partial fraction decomposition to decompose the integrand into simple fractions as

$$\frac{1}{z^2 - 1} = \frac{1}{2(z-1)} - \frac{1}{2(z+1)}.$$

This is something that we will use again and again in different contexts! The contour integral becomes

$$I = \oint_C \frac{dz}{2(z-1)} - \oint_C \frac{dz}{2(z+1)}.$$

The first integral vanishes since its integrand is analytic everywhere inside C. For the second integral, since $z = -1$ is inside C, we apply the Cauchy's integral formula. Therefore,

$$I = -\frac{1}{2}(2\pi i) = -\pi i.$$

Alternatively, we could rewrite the integrand to isolate the singularity which is inside the contour, namely

$$I = \oint_C \frac{f(z)}{z+1} dz,$$

where

$$f(z) = \frac{1}{z-1} dz$$

is analytic inside the contour. By Cauchy's theorem, it follows that the integral equals

$$I = 2\pi i f(1) = -\pi i.$$

Example 31 Compute the contour integral

$$I = \oint_C \frac{\sin(3z)}{2z - \pi} dz = \frac{1}{2} \oint_C \frac{\sin(3z)}{z - \pi/2} s$$

where C is:
(a) $|z| = 1$ oriented counterclockwise.
(b) $|z| = 2$ oriented counterclockwise.

Solution: Let us first cast the integral so that we can readily apply the Cauchy's integral formula, namely as

$$I = \frac{1}{2} \oint_C \frac{\sin(3z)}{z - \pi/2} dz,$$

4 Lecture 4: Complex Integration

from which we see that $f(z) = \sin(3z)$. Now, we need to find if $z_0 = \pi/2$ is inside the integration domain or not.

(a) $z_0 = \pi/2$ is outside the unit disk centered at 0. For the contour integral over this unit circle vanishes by Cauchy's theorem:

$$I = 0$$

(b) In this case, $z_0 = \pi/2$ is inside the disk of radius 2 enclosed by C given by $|z| = 2$. Thus, by the Cauchy's integral formula

$$I = \frac{1}{2} 2\pi i f(\pi/2) = i\pi \sin(3\pi/2) = -i\pi.$$

Example 32 Compute the contour integral

$$I = \oint_{C_1} \frac{z^3 e^{2z}}{2z + i} dz$$

where C is the unit circle centered at origin: $|z| = 1$ traversed clockwise.

Solution: First, we rewrite the integral as

$$I = -\frac{1}{2} \oint_C \frac{z^3 e^{2z}}{z + i/2} dz, \quad C : |z| = 1 \text{ counter clockwise}$$

From the integrand, we see that $f(z) = z^3 e^{2z}$ is analytic and $z_0 = -i/2$ is inside C. Hence, by Cauchy's line integral, the integral is evaluated to

$$I = -\frac{1}{2} 2\pi i f(-i/2) = \frac{\pi}{8} e^{-i}.$$

Example 33 (*See Fig.* 9) Compute the contour integral

$$I = \oint_C \frac{\cos z}{z^2 - 4} dz$$

where C is the rectangular contour with corners $i, -i, 3-i, 3+i$ traversed in the counter-clockwise direction.

Solution: The integrand has isolated singularities at ± 2, but only $z_0 = 2$ is inside the contour. Thus, we can use the Cauchy's integral formula with $f(z) = \frac{\cos(z)}{z+2}$ which is analytic inside the rectangle C. Thus,

$$I = \oint_C \frac{f(z)}{z - 2} dz,$$

Fig. 9 Rectangular contour from Exercise 33. The singular points are marked by the red dots

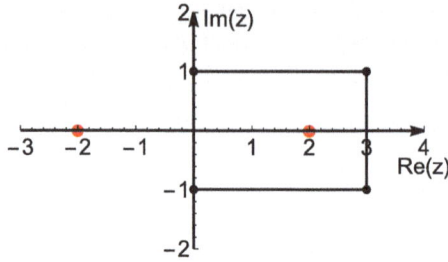

which results in

$$I = 2\pi i f(2) = 2\pi i \frac{\cos(2)}{4} = i\pi \cos(2).$$

5 Lecture 5: Generalized Cauchy's Integral Formula

In this lecture, we introduce the generalized Cauchy's integral formula, which offers a contour integral representation for the n'th derivatives of a complex function. We also discuss Cauchy's inequality, which is important in proving the convergence of power series expansions.

5.1 Generalized Cauchy's Integral Theorem

Differential Form of Cauchy's Integral Formula: If $f(z)$ is analytic on and inside a simple contour C, and z_0 is a point inside C, then the value of the n'th derivative $f^{(n)} = d^n f/dz^n$ at $z = z_0$ has an integral form given by

$$\boxed{f^{(n)}(z_0) = \frac{n!}{2\pi i} \oint_C \frac{f(z)}{(z-z_0)^{n+1}} dz, \qquad z_0 \text{ inside } C} \tag{24}$$

The function $f(z)$ that is analytic inside C is differentiable to any order.

Proof This can be shown by differentiating recursively Cauchy's integral formula:

$$f'(z_0) = \frac{1}{2\pi i} \oint_C \frac{f(z)}{(z-z_0)^2} dz$$

$$f^{(2)}(z_0) = \frac{2!}{2\pi i} \oint_C \frac{f(z)}{(z-z_0)^3} dz$$

...

5 Lecture 5: Generalized Cauchy's Integral Formula

Fig. 10 Contour $|z| = 3$ for Exercise 34

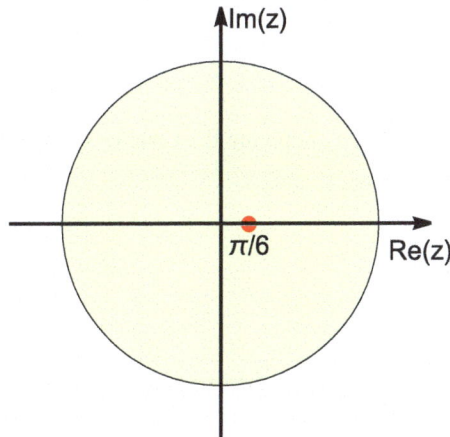

$$f^{(n)}(z_0) = \frac{n!}{2\pi i} \oint_C \frac{f(z)}{(z - z_0)^{n+1}} dz$$

Example 34 (*See Fig.* 10) Evaluate

$$I = \oint_C dz \frac{2\sin(2z)}{(6z - \pi)^3},$$

where C is the circle $|z| = 3$.

Solution: We notice that $z_0 = \pi/6$ is inside the circle at 0 and of radius 3. Furthermore, $f(z) = \frac{2\sin(2z)}{6^3}$ is analytic inside C. Hence, the conditions of the Cauchy's integral formula apply, and we get for $n = 2$

$$I = \oint_C dz \frac{2\sin(2z)}{(6z - \pi)^3} = \frac{2\pi i}{2!} f^{(2)}(\pi/6)$$

$$= -\pi i \frac{2^3 \sin(\pi/3)}{6^3} = -i \frac{\pi}{3^3} \frac{\sqrt{3}}{2}$$

Example 35 Evaluate

$$I = \oint_C dz \frac{e^{iz}}{(z - i\pi)^3},$$

where C is the circle $|z - i\pi| = 1$.

Solution: Let us denote by $z_0 = i\pi$ and $f(z) = e^{iz}$ which is analytic inside C. Its first two derivatives are

$$f' = ie^{iz}, \qquad f^{(2)} = -e^{iz}.$$

Hence, by applying the Cauchy's integral formula for $n = 2$, we find

$$I = \frac{2\pi i}{2!} f^{(2)}(z_0)$$
$$= -i\pi e^{-\pi} = -i\pi.$$

5.2 Cauchy's Inequality

The Cauchy's inequality is particularly useful in proving the convergence of power series.

Triangle Inequality: Let $f(z)$ be bounded on a circle C of radius R centered at z_0. i.e. $|f(z)| \leq M$ for all z on C. Then, the contour integral over C is also upper-bounded

$$\boxed{\left| \oint_C f(z) dz \right| \leq 2\pi R M} \qquad (25)$$

Proof

$$\left| \oint_C f(z) dz \right| \leq \oint_{|z-z_0|=R} |f(z)||dz| \leq M \oint_{|z-z_0|=R} |dz|$$

Using the parameterization $z = z_0 + Re^{i\theta}$, with $0 \leq \theta < 2\pi$ and $|dz| = |iRe^{i\theta} d\theta| = R d\theta$, we have that

$$\oint_{|z-z_0|=R} |dz| = R \int_0^{2\pi} d\theta = 2\pi R.$$

Thus,

$$\left| \oint_C f(z) dz \right| \leq 2\pi R M.$$

5 Lecture 5: Generalized Cauchy's Integral Formula

Example 36 Find an upper bound for the absolute value of

$$I = \oint_C \frac{e^z}{z^2 - 1} dz,$$

where the contour is the circle of radius 2 centered at the origin, i.e. $|z| = 2$, traversed in the counterclockwise direction.

Solution: The contour C corresponds to $|z| = 2$ with the circumference of length $2\pi R = 4\pi$. Next, we need to determine the upper bound M of $|f(z)|$.

$$|e^z| = |e^x e^{iy}| = e^x \leq e^2$$

We use the triangle inequality

$$|z_1 + z_2| \leq |z_1| + |z_2|$$

adapted to our case as

$$|(z^2 - 1) + 1| \leq |z^2 - 1| + 1 \rightarrow |z^2 - 1| \geq |z^2| - 1$$

Combining this with the upper bound for the magnitude of z, we have that

$$|z^2 - 1| \geq |z|^2 - 1 = R^2 - 1 \geq 3$$

Hence,

$$|f(z)| = \frac{|e^z|}{|z^2 - 1|} \leq \frac{e^2}{3}$$

and the integral is upper bounded in magnitude by

$$|I| \leq \frac{4\pi e^2}{3}$$

Let us evaluate the integral using Cauchy's formula. Notice that the disk of radius 2 includes both singular points $z = \pm 1$. To deal with them separately, we use the partial fraction decomposition and arrive at

$$I = \oint_{|z|=2} \frac{e^z}{z^2 - 1} dz,$$

$$= \frac{1}{2} \oint_{|z|=2} \frac{e^z}{z - 1} dz - \frac{1}{2} \oint_{|z|=2} \frac{e^z}{z + 1} dz,$$

$$= \pi i (e^1 - e^{-1}).$$

Thus,
$$|I| = \pi e^1 (1 - e^{-2}) < \frac{4\pi e^2}{3}.$$

Generalized Cauchy's Inequality: Let $f(z)$ be analytic on and inside a circle C of radius R centered at z_0. If $f(z)$ is bounded, i.e. $|f(z)| \leq M$ for all z on C, then its derivatives are upper-bounded in the center of the disk by

$$\boxed{|f^{(n)}(z_0)| \leq \frac{n!M}{R^n}}. \tag{26}$$

This is a useful inequality in proving important theorems. We will use it later in the proof of the Taylor expansion.

Proof When $f(z)$ is analytic inside the disk centered at z_0 and of radius R, then it is also upper-bounded, i.e. there exits $M > 0$ such that $|f(z)| \leq M$. Let us start from the generalized Cauchy's integral formula

$$|f^{(n)}(z_0)| = \left| \frac{n!}{2\pi i} \oint_{|z-z_0|=R} \frac{f(z)}{(z-z_0)^{n+1}} dz \right|$$

$$\leq \frac{n!}{2\pi} \oint_{|z-z_0|=R} \frac{|f(z)|}{|z-z_0|^{n+1}} |dz|$$

$$\leq \frac{n!}{2\pi} \frac{M}{R^{n+1}} \oint_{|z-z_0|=R} |dz|$$

Using the parameterization $z = z_0 + Re^{i\theta}$, with $0 \leq \theta < 2\pi$ and $|dz| = |iRe^{i\theta} d\theta| = R d\theta$, we have that $\oint_{|z-z_0|=R} |dz| = 2\pi R$.

$$|f^{(n)}(z_0)| \leq \frac{n!M}{R^n}$$

We may notice that this upper bound may not apply for derivatives that are evaluated at other points inside the disk other than its center z_0.

Example 37 Let us consider the function $f(z) = z^2 e^z$ from previous examples, and evaluate the upper found of $f^{(2)}(0)$ at the center of the circle $|z| = 2$.

Solution: First, let us determine the upper bound M of $f(z)$ on the circle $|z| = 2$:

$$|f(z)| = |z|^2 |e^z| = |z|^2 e^x \leq 2^2 e^2$$

Thus, $M = 2^2 e^2$ and $R = 2$. According to the generalized Cauchy's inequality:

$$|f^{(2)}(0)| \leq 2e^2$$

Theorem 6 (Liouville's theorem) *If $f(z)$ is analytic and bounded in the entire complex plane, then $f(z)$ is constant.*

Proof We want to prove that $f(z)$ is constant hence $f'(z) = 0$. Since $f(z)$ is analytic and bounded by M everywhere i.e. $|f(z)| < M$ for every $z \in \mathbb{C}$. This implies that $f(z)$ is analytic in a disk around an *arbitrary* point $z_0 \in \mathbb{C}$ ($|z - z_0| \leq R$). By Cauchy's inequality, there exists an upper bound M_R for $f(z)$ in this disk of radius R, such that the derivative is bounded by:

$$\boxed{|f'(z_0)| \leq \frac{M_R}{R}}$$

Since the disk is of arbitrary size R and center z_0, the inequality is also satisfied in the limit of $R \to \infty$ and for any z_0. But, since the absolute value of positive defined, it means that

$$\boxed{0 \leq |f'(z_0)| \leq 0 \Rightarrow f'(z_0) = 0, \forall z_0 \Rightarrow f(z) = constant.}$$

The immediate consequence of this is that harmonic function, i.e. $u = Re(f)$ and $v = Im(f)$ that are bounded in the plane are constant functions. In other words, the bounded solution of the Laplace equation over the 2D plane is a constant.

6 Lecture 6: Taylor Expansion

We introduce the complex power series expansions of complex functions. This is particularly useful as it represents a complex function in a form that reveals the isolated singularities in a given integration domain. In this lecture, we focus on the Taylor expansion of analytic functions and demonstrate its application through several examples.

6.1 Taylor Expansion

Theorem 7 *If $f(z)$ is analytic at z_0 then it can be represented as a complex power series also known as Taylor expansion*

$$\boxed{f(z) = \sum_{n=0}^{\infty} a_n (z - z_0)^n} \qquad (27)$$

where the complex coefficients are uniquely defined by the derivatives of $f(z)$ evaluated at z_0,

$$\boxed{a_n = \frac{f^{(n)}(z_0)}{n!} = \frac{1}{2\pi i} \oint \frac{f(w) dw}{(w - z_0)^{n+1}}} \qquad (28)$$

The complex function $f(z)$ has a **unique** Taylor series expansion around a regular point z_0. The disk of convergence is centered at z_0 and extends to the nearest singularity.

Proof We make use of the geometric series expansion

$$\boxed{\frac{1}{1-r} = \sum_{n=0}^{\infty} r^n, \text{ for } r < 1.} \qquad (29)$$

First thing to show is that the Taylor series is convergent for any z in the region centered at z_0 where $f(z)$ is analytic. For a compact analytic region centered at z_0 and of radius R, we have that $f(z)$ is bounded, i.e. there exits an $M > 0$ such that

$$|f(z)| \leq M.$$

Thus, by the Cauchy's inequality, it follows that its derivatives at z_0 are upper-bounded by

$$|f^{(n)}(z_0)| \leq \frac{n!M}{R^n} \rightarrow |a_n| \leq \frac{M}{R^n}.$$

It follows that each term in the power series is bounded by

$$|a_n (z - z_0)^n| = |a_n||z - z_0|^n \leq M \frac{r^n}{R^n}, \qquad r = |z - z_0|$$

These terms correspond to those in the geometric series $\sum_n (r/R)^n$, which converges to $(1 - \frac{r}{R})^{-1}$ for $r < R$. Since $r = |z - z_0|$ represents the radius of a small region within the analytic domain of size R, the condition $r < R$ is inherently satisfied. Therefore, each term in the Taylor series is bounded by the corresponding term in the geometric series, implying that the Taylor series is also convergent (*see Fig. 11*).

The second thing is to show is that the coefficients a_n of the power series are determined by the derivatives of $f(z)$. We start from the Cauchy's integral formula

$$f(z) = \frac{1}{2\pi i} \oint_{C_R} \frac{f(w)}{w - z} dw$$

6 Lecture 6: Taylor Expansion

Fig. 11 Disk of convergence for Taylor expansion at z_0

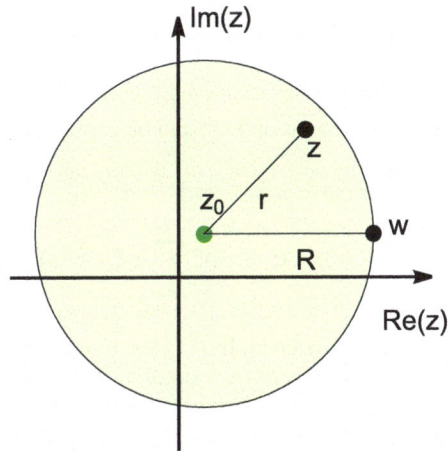

where the contour is the circle centered at z_0 and of radius R.

$$f(z) = \frac{1}{2\pi i} \oint_{C_R} \frac{f(w)}{(w - z_0) - (z - z_0)} dw$$

$$= \frac{1}{2\pi i} \oint_{C_R} \frac{f(w)}{w - z_0} \left(1 - \frac{z - z_0}{w - z_0}\right)^{-1} dw.$$

By expanding the simple fraction into its corresponding geometric series for $|z - z_0| < |w - z_0|$, and rearranging terms we have that

$$f(z) = \frac{1}{2\pi i} \oint_{C_R} \frac{f(w)}{w - z_0} \sum_{n=0}^{\infty} \left(\frac{z - z_0}{w - z_0}\right)^n dw, \quad \frac{|z - z_0|}{|w - z_0|} = \frac{r}{R} < 1$$

$$= \sum_{n=0}^{\infty} (z - z_0)^n \frac{1}{2\pi i} \oint_{C_R} \frac{f(w)}{(w - z_0)^{n+1}} dw$$

$$= \sum_{n=0}^{\infty} \frac{f^{(n)}(z_0)}{n!} (z - z_0)^n$$

6.1.1 Definitions: Zeros

Suppose $f(z)$ is analytic on a disk $|z - z_0| < R$ such that it has a unique Taylor expansion around z_0.

A zero: If $f(z_0) = 0$, then z_0 is a *zero* of $f(z)$ when $f(z_0) = 0$ and not all the a_n are zero.

A zero of order n: z_0 is a *zero of order n* of $f(z)$ when $a_m = 0$ for $m < n$. Then, the Taylor expansion of $f(z)$ can be expressed as

$$f(z) = a_n(z - z_0)^n + a_{n+1}(z - z_0)^{n+1} + + a_{n+2}(z - z_0)^{n+2} + \cdots$$

$$f(z) = (z - z_0)^n \sum_{k=n}^{\infty} a_k (z - z_0)^{k-n}.$$

Zeros are isolated. If $f(z)$ has more than one zero, we can always isolate them by enclosing each zero by a small disk that does not contain any other zeros.

Example 38 Find the Taylor series expansion of $f(z) = e^{2z}$ around a point z_0 in the complex plane.

Solution: Since the exponential is analytic everywhere in the complex plane, it has a well-defined Taylor expansion around any point z_0. To determine the Taylor coefficients, we compute:

$$f^{(n)}(z) = \frac{d^n}{dz^n} e^{2z} = 2^n e^{2z}$$

Hence,

$$e^{2z} = e^{2z_0} \sum_n \frac{2^n}{n!} (z - z_0)^n$$

Example 39 Find the Taylor expansion of $f(z) = \frac{\sin(z)}{z}$ at $z_0 = 0$.

Solution: Apparently, there seems to be a problem at $z = 0$ because of the singular behavior of $1/z$. However, on a closer inspection, we may recognise that this is removed by the first order term in the Taylor expansion of $\sin(z)$ which is analytic at $z_0 = 0$. Thus, $\sin(z)/z$ is analytic and has a Taylor expansion at $z_0 = 0$. We call such a point a *removable singularity*.

To check this let us first look at the Taylor expansion of $\sin(z)$ at $z_0 = 0$. The corresponding coefficients are

$$a_0 = \sin(z)|_{z=0} = 0 \tag{30}$$

$$a_1 = \cos(z)|_{z=0} = 1 \tag{31}$$

$$a_2 = \frac{1}{2!}(-1)\sin(z)|_{z=0} = 0 \tag{32}$$

$$a_3 = \frac{1}{3!}(-1)\cos(z)|_{z=0} = \frac{1}{3!}(-1) \tag{33}$$

$$\ldots \quad (34)$$

$$a_{2k+1} = \frac{1}{(2k+1)!}(-1)^k \quad (35)$$

Thus, its Taylor series is given as

$$\sin(z) = \sum_{k=0}^{\infty} \frac{1}{(2k+1)!}(-1)^k z^{2k+1}$$

from which it follows that

$$f(z) = \frac{\sin(z)}{z} = \sum_{k=0}^{\infty} \frac{1}{(2k+1)!}(-1)^k z^{2k}$$

Furthermore, we observe that the disk of convergence spans the whole complex plane.

Example 40 (*See Fig.* 12) Find the Taylor expansion of the principal value of the logarithm $f(z) = \ln(z)$ around the point $z_0 = 1$ where it is analytic.

Solution: Recall that the normal form of the logarithm is

$$\mathrm{Ln}(z) = \mathrm{Ln}(re^{i\theta}) = \ln(r) + i\theta, \ -\pi < \theta \leq \pi.$$

We notice that $\mathrm{Ln}(z)$ is not analytic at the origin. This singular point is however not isolated, and this is because of the sudden jump in the phase along the real non-positive axis. The origin is called the **branch point** and the line of phase discontinuity is the **branch cut**. The branch point is uniquely defined, however the branch cut depends on the principal value interval of the argument.

Fig. 12 Disk of convergence for Taylor expansion of $\mathrm{Ln}(z)$ around $z = 1$

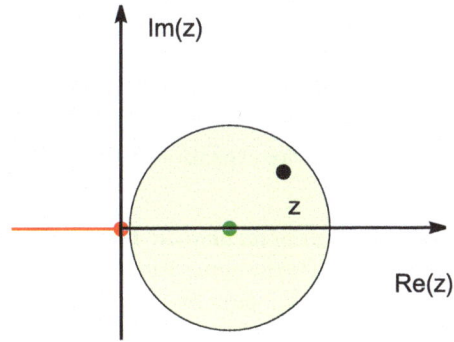

Now, we are looking at a point on the positive real axis. Hence, the disk of convergence, being the circular disk touching the nearest singularity, will be $|z - 1| < 1$. Using that

$$f^{(1)}(z) = \frac{d\mathrm{Ln}(z)}{dz} = \frac{1}{z}$$

we can evaluate the other derivatives by successive differentiation

$$f^{(2)}(z) = \frac{d}{dz}\frac{1}{z} = (-1)\frac{1}{z^2}$$

$$\ldots$$

$$f^{(n)}(z) = \frac{d^n}{dz^n}\mathrm{Ln}(z) = (-1)^{n-1}(n-1)!\frac{1}{z^n}$$

Hence,

$$\mathrm{Ln}(z) = \sum_{n=0}^{\infty} \frac{f^{(n)}(1)}{n!}(z-1)^n$$

$$= \ln(1) + \sum_{n=1}^{\infty} \frac{(-1)^{n-1}}{n}(z-1)^n$$

$$= \sum_{n=1}^{\infty} \frac{(-1)^{n-1}}{n}(z-1)^n$$

Let us see what happens when we differentiate the series expansion of the logarithm. Then, we get that

$$\frac{d}{dz}\sum_{n=1}^{\infty} \frac{(-1)^{n-1}}{n}(z-1)^n = \sum_{n=1}^{\infty}(-1)^{n-1}(z-1)^{n-1}$$

$$= \sum_{n=1}^{\infty}(-1)^{n-1}(-1)^{n-1}(1-z)^{n-1} = \sum_{n=0}^{\infty}(1-z)^n$$

$$= \frac{1}{1-(z-1)} = \frac{1}{z}$$

which is the Taylor expansion of $1/z$ for the same convergent disk $|z - 1| < 1$ as the $\mathrm{Ln}(z)$.

The derivatives of an analytic function are also analytic. If a function $f(z)$ is analytic at z_0, the Taylor series for $f'(z)$ around z_0 can be derived by differentiating the Taylor series for f term by term. Both Taylor expansions will share the same disk of convergence.

7 Lecture 7: Laurent Expansion

In this lecture, we introduce the general representation of $f(z)$ as a power series with both positive and negative powers, called the Laurent series expansion. The function $f(z)$ may have different Laurent expansions at z_0 for different domains occupied by z. In this respect, the Taylor expansion is a particular form of Laurent expansion where z_0 is a regular point and the center of the analytic expansion domain.

7.1 Laurent Expansion

Theorem 8 (Laurent's theorem) *Let C_1 and C_2 be two circles of radius R_1 and $R_1 > R_2$, respectively and centered at the same point z_0. When, $f(z)$ is analytic in the annulus $R_2 < |z - z_0| < R_1$, it can be expanded in a Laurent series:*

$$f(z) = \sum_{n=0}^{\infty} a_n (z-z_0)^n + \sum_{n=1}^{\infty} b_n \frac{1}{(z-z_0)^n} \tag{36}$$

with the coefficients uniquely determined by $f(z)$ and the analytic domain where the function is expanded:

$$a_n = \frac{1}{2\pi i} \oint_C \frac{f(z)}{(z-z_0)^{n+1}} dz, \quad b_n = \frac{1}{2\pi i} \oint_C \frac{f(z)}{(z-z_0)^{-n+1}} dz \tag{37}$$

*where C is an **arbitrary** closed contour inside the annulus, and encircling z_0 (does not have to be the same contour for both integrals).*

Regular Part: $\sum_{n=0}^{\infty} a_n (z-z_0)^n$ converges for $|z-z_0| < R_1$. This series is called the **analytic or regular part** of the Laurent series.

Singular Part: $\sum_{n=1}^{\infty} b_n (z-z_0)^{-n}$ is convergent for $|z-z_0| > R_2$. This series is called the **principal or singular part** of the Laurent series.

Both the regular part and the singular part of the Laurent series are convergent inside the annulus $R_2 < r < R_1$, where $f(z)$ is analytic.

Alternatively, we may write the Laurent expansion in this compact form

$$f(z) = \sum_{n=-\infty}^{\infty} c_n (z-z_0)^n \tag{38}$$

Fig. 13 Contour circumventing z_0 and enclosing the region where $f(z)$ is analytic

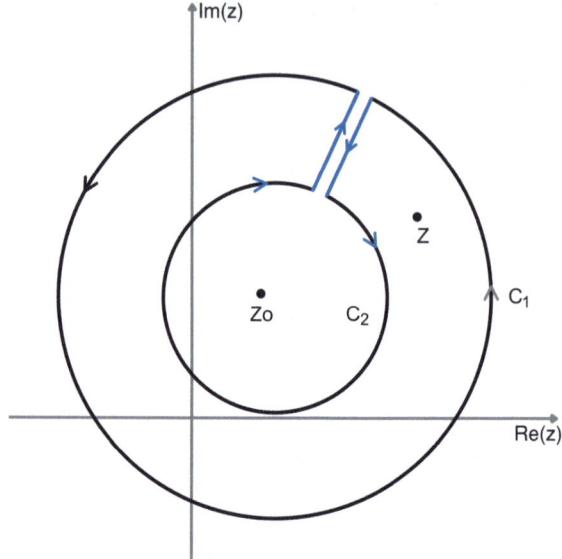

where

$$c_n = \frac{1}{2\pi i} \oint_C \frac{f(z)}{(z-z_0)^{n+1}} dz. \qquad (39)$$

Proof (*see Fig.* 13) The point z_0 could either regular or singular. However, $f(z)$ needs to be analytic in the domain where it is expanded to ensure that the Laurent series is convergent. This convergence domain is determined by the annulus $R_2 < |z - z_0| < R_1$. We encircle the point z by a contour joining C_2 and C_1 as illustrated in Fig. 13. Contour C_1 has counterclockwise (positive) orientation, while the contour C_2 has clockwise (negative) orientation and the paths ($\pm\Gamma$) bridging these two contours have opposite orientations. Since $f(z)$ is analytic inside this closed path, we can use Cauchy's integral formula to represent $f(z)$ as

$$f(z) = \frac{1}{2\pi i} \oint_C \frac{f(w)}{w-z} dw$$

$$= \frac{1}{2\pi i} \oint_{C_1} \frac{f(w)}{w-z} dw - \frac{1}{2\pi i} \oint_{C_2} \frac{f(w)}{w-z} dw$$

For the first integral, we have that z is inside C_1 hence $\left|\frac{z-z_0}{w-z_0}\right| < 1$. This implies that the integral over this contour becomes

7 Lecture 7: Laurent Expansion

$$\frac{1}{2\pi i}\oint_{C_1}\frac{f(w)}{w-z}dw = \frac{1}{2\pi i}\oint_{C_1}\frac{f(w)}{(w-z_0)-(z-z_0)}dw \qquad (40)$$

$$= \frac{1}{2\pi i}\oint_{C_1}\frac{f(w)}{w-z_0}\frac{1}{1-\frac{z-z_0}{w-z_0}}dw \qquad (41)$$

Notice that since $|z - z_0| < |w - z_0|$, the simple fraction can be expanded as a geometric series for every w on C_1, thus

$$\frac{1}{2\pi i}\oint_{C_1}\frac{f(w)}{w-z}dw = \frac{1}{2\pi i}\oint_{C_1}\frac{f(w)}{w-z_0}\sum_{n=0}^{\infty}\left(\frac{z-z_0}{w-z_0}\right)^n dw$$

$$= \sum_{n=0}^{\infty}\left(\frac{1}{2\pi i}\oint_{C_1}\frac{f(w)dw}{(w-z_0)^{n+1}}\right)(z-z_0)^n$$

$$= \sum_{n=0}^{\infty}a_n(z-z_0)^n. \qquad (42)$$

What we have shown is that the first integral reduces to the regular series which is convergent inside C_1. Notice that the coefficients a_n can be related to the derivatives of $f(z)$ through the Cauchy's integral formula only when $f(z)$ is analytic at z_0!

Now, for the second contour integral, z is outside the contour C_2 hence $\left|\frac{w-z_0}{z-z_0}\right| < 1$. Thus, the contour integral can be rewritten as

$$\frac{1}{2\pi i}\oint_{C_2}\frac{f(w)}{w-z}dw = \frac{1}{2\pi i}\oint_{C_2}\frac{f(w)}{(w-z_0)-(z-z_0)}dw$$

$$= -\frac{1}{2\pi i}\oint_{C_2}\frac{f(w)}{z-z_0}\frac{1}{1-\frac{w-z_0}{z-z_0}}dw. \qquad (43)$$

For this circle, the distance from z to z_0 is larger than its radius, i.e. $|z-z_0| > |w-z_0|$, thus we can use the geometric series expansion in this form

$$\frac{1}{2\pi i}\oint_{C_2}\frac{f(w)}{w-z}dw = -\frac{1}{2\pi i}\oint_{C_2}\frac{f(w)}{z-z_0}\sum_{n=0}^{\infty}\left(\frac{w-z_0}{z-z_0}\right)^n dw$$

$$= -\sum_{n=0}^{\infty}\left(\frac{1}{2\pi i}\oint_{C_2}\frac{f(w)dw}{(w-z_0)^{-n}}\right)\frac{1}{(z-z_0)^{n+1}}$$

$$= -\sum_{n=1}^{\infty}\left(\frac{1}{2\pi i}\oint_{C_2}\frac{f(w)dw}{(w-z_0)^{-n+1}}\right)\frac{1}{(z-z_0)^n}$$

$$= -\sum_{n=1}^{\infty} b_n \frac{1}{(z-z_0)^n}. \tag{44}$$

This second integral reduces to the singular part of the Laurent series which is convergent for z being outside C_1 and inside C_2.

The domain where both series are convergent simultaneously corresponds to the annulus $R_2 < |z - z_0| < R_1$. This is the convergence domain for the Laurent expansion at z_0,

$$f(z) = \frac{1}{2\pi i} \oint_C \frac{f(w)}{w-z} dw$$

$$= \frac{1}{2\pi i} \oint_{C_1} \frac{f(w)}{w-z} dw - \frac{1}{2\pi i} \oint_{C_2} \frac{f(w)}{w-z} dw$$

$$= \sum_{n=0}^{\infty} a_n (z-z_0)^n + \sum_{n=1}^{\infty} b_n \frac{1}{(z-z_0)^n}$$

7.2 Definitions: Poles and Residue

Definition 9 (*Regular point*) When $b_n = 0$ for all n's, then $f(z)$ is analytic at z_0, and z_0 is a *regular point*. Then, the domain where $f(z)$ is expanded can also include z_0.

Definition 10 (*Simple Pole*) If $b_1 \neq 0$ and $b_k = 0$, $k > 1$, then z_0 is a *simple pole*

$$\frac{b_1}{z-z_0} + \sum_n a_n (z-z_0)^n$$

Definition 11 (*Pole of order n*) If $b_n \neq 0$ for some n, but $b_k = 0$ for all $k > n$, then $f(z)$ is said to have a *pole of order n* at z_0.

$$\frac{b_1}{z-z_0} + \cdots \frac{b_n}{(z-z_0)^n} + \sum_n a_n (z-z_0)^n$$

Definition 12 (*Residue*) When $f(z)$ is expanded at z_0 in a domain centered at z_0, then the coefficient b_1 of the Laurent expansion is called the *residue* of $f(z)$ at z_0: $Res(f, z_0) = b_1$. When z_0 is a regular point, the function is analytic at z_0 and is Taylor expanded at z_0. Thus, the residue at a regular point is zero. However, when z_0 is an isolated singularity, the Laurent expansion in the disk punctuated at z_0, will pick up singular terms and the residue can be non-zero.

7 Lecture 7: Laurent Expansion

7.3 Examples

Example 41 Find the Laurent expansion of

$$f(z) = \frac{1}{(z-1)(z-2)}$$

at $z_0 = 0$ for $|z| < 1$. What is its residue at $z_0 = 0$?

Solution:

Method I: Determine the c_n Coefficients We aim to determine the Laurent expansion around z_0 in this form

$$f(z) = \frac{1}{(z-1)(z-2)} = \sum_{n=-\infty}^{\infty} c_n z^n \tag{45}$$

where

$$c_n = \frac{1}{2\pi i} \oint_C \frac{f(z)}{z^{n+1}} dz. \tag{46}$$

for a contour that is inside the disk $|z| < 1$. We notice that $f(z)$ is analytic inside the contour C. Since $f(z)$ is analytic at $z_0 = 0$, we can relate c_n with derivatives of $f(z)$ for $n \geq 0$:

$$c_n = \frac{f^{(n)}(0)}{n!}, \quad n \geq 0. \tag{47}$$

For $n < 0$, the integrand $f(z)z^{-n-1}$ is analytic and, by the Cauchy's theorem, it follows that the contour integral vanishes, hence

$$c_n = 0, \quad n \leq -1. \tag{48}$$

Thus, the singular part of the Laurent series vanishes and we are left with the Taylor series expansion at the regular point $z_0 = 0$:

$$f(z) = \frac{1}{(z-1)(z-2)} = \sum_{n=0}^{\infty} \frac{f^{(n)}(0)}{n!} z^n \tag{49}$$

which is convergent for $|z| < 1$. Now we need to determine the derivatives:

$$\frac{d^n}{dz^n}\left(\frac{1}{(z-2)(z-1)}\right) = \frac{d^n}{dz^n} \frac{1}{(z-2)} - \frac{d^n}{dz^n} \frac{1}{(z-1)}$$

$$= \frac{(-1)^n n!}{(z-2)^{n+1}} - \frac{(-1)^n n!}{(z-1)^{n+1}}$$

Hence,

$$f^{(n)}(0) = (-1)^{n-n-1} n! 2^{-n-1} - (-1)^{n-n-1} n! = n!(1 - 2^{-n-1})$$

such that the series expansion becomes

$$f(z) = \frac{1}{(z-1)(z-2)} = \sum_{n=0}^{\infty} (1 - 2^{-n-1}) z^n \qquad (50)$$

Method II: Partial Fraction Decomposition We use the partial fraction decomposition

$$f(z) = \frac{1}{z-2} - \frac{1}{z-1},$$

from which we locate the isolated singularities $z = 1$ and $z = 2$ (see Fig. 14). The disk with $|z| < 1$ does not contain these singular points, i.e. $f(z)$ is analytic inside it. We fulfill the conditions for a Taylor expansion, hence we are expecting that the Laurent series will contain only the regular part.

We expand the simple fractions at $z_0 = 0$ for their corresponding disks of convergence centered at $z_0 = 0$. For $(z-2)^{-1}$ this corresponds to $|z| < 2$:

$$\frac{1}{z-2} = -\frac{1}{2} \frac{1}{1-z/2} = -\frac{1}{2} \sum_{n=0}^{\infty} \left(\frac{z}{2}\right)^n \qquad (51)$$

The disk of convergence $|z| < 2$ includes the unit disk.

Fig. 14 Expansion domain for Example 41

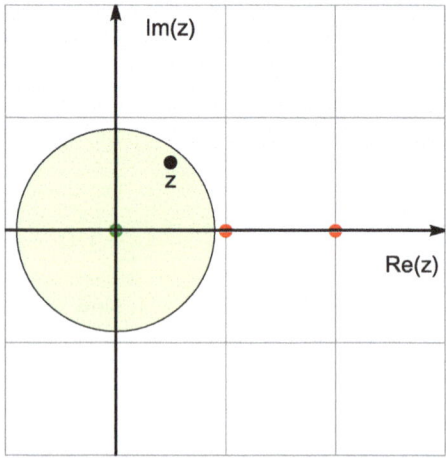

7 Lecture 7: Laurent Expansion

Similarly, for $(z-1)^{-1}$ we use the convergence disk $|z| < 1$:

$$\frac{1}{z-1} = -\frac{1}{1-z} = -\sum_{n=0}^{\infty} (z)^n. \tag{52}$$

Hence $f(z)$ is expanded as

$$\frac{1}{(z-2)(z-1)} = \sum_{n=0}^{\infty} \left(1 - \frac{1}{2^{n+1}}\right) z^n,$$

which coincides with the Taylor series. Since $b_1 = 0$, it follows that the residue $Res(f, 0) = 0$.

This is the more trivial case when z_0 is a regular point *inside* the expansion domain, and thus we obtain the Taylor expansion. Now, we will consider the same function and same $z_0 = 0$, but take different expansion domains which do not include z_0 (Fig. 15).

Example 42 Find the Laurent expansion of

$$f(z) = \frac{1}{(z-1)(z-2)}$$

at $z_0 = 0$ for $|z| > 2$.

Solution: In this case, the domain of expansion is detached from $z_0 = 0$ and corresponds to the outside of the disk $|z| > 2$ (see Fig. 14). The geometric series used previously are not convergent in this domain. So, we need to adapt the simple fractions. This is done in the following way:

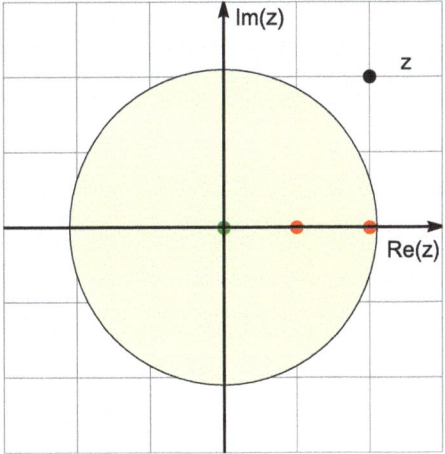

Fig. 15 Expansion domain for Example 42

$$\frac{1}{z-1} = \frac{1}{z}\frac{1}{(1-1/z)} = \frac{1}{z}\sum_{n=0}^{\infty}\left(\frac{1}{z}\right)^n = \sum_{n=0}^{\infty}\frac{1}{z^{n+1}}, \quad |z|>1 \qquad (53)$$

$$\frac{1}{z-2} = \frac{1}{z}\frac{1}{(1-2/z)} = \frac{1}{z}\sum_{n=0}^{\infty}\left(\frac{2}{z}\right)^n = \sum_{n=0}^{\infty}\frac{2^n}{z^{n+1}}, \quad |z|>2 \qquad (54)$$

Collecting the two series, we find that

$$\frac{1}{(z-1)(z-2)} = \sum_{n=0}^{\infty}\frac{2^n-1}{z^{n+1}}, \quad |z|>2$$

which contains only the singular part of Laurent expansion (Fig. 16).

Example 43 Find the Laurent expansion of

$$f(z) = \frac{1}{(z-1)(z-2)}$$

at $z_0 = 0$ for $1 < |z| < 2$.

Solution: In this case, z is inside the annulus determined by the two concentric circles with center at the origin (see Fig. 14). Equations 53 and 52 are the convergent geometric series for this domain, thus the Laurent expansion of $f(z)$ can be written as

$$\frac{1}{(z-1)(z-2)} = -\sum_{n=0}^{\infty}\frac{1}{2^{n+1}}z^n - \sum_{n=0}^{\infty}\frac{1}{z^{n+1}}, \quad 1<|z|<2$$

Fig. 16 Expansion domain for Example 43

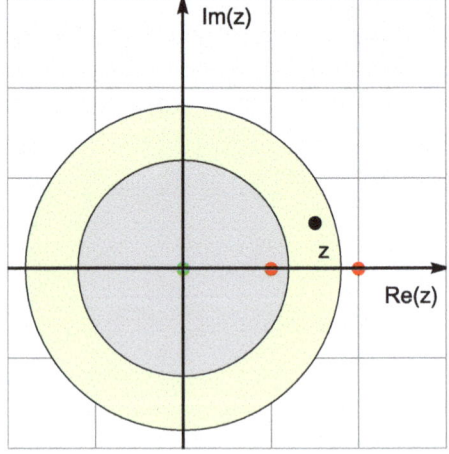

7 Lecture 7: Laurent Expansion

Fig. 17 Expansion domain for Example 44

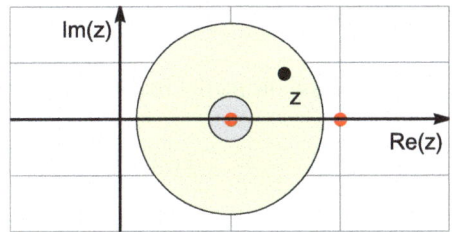

which contains both the regular and the singular parts of Laurent expansion (Fig. 17).

Example 44 Find the Laurent expansion of

$$f(z) = \frac{1}{(z-1)(z-2)}$$

at $z_0 = 1$ for $0 < |z-1| < 1$ (disk punctuated at $z_0 = 1$ and of radius smaller than 1).

Solution: $(z-2)^{-1}$ is analytic inside this domain including $z_0 = 1$, thus we can Taylor expand it at $z_0 = 1$. Hence, the Laurent expansion of $f(z)$ at $z_0 = 1$ is

$$\frac{1}{(z-1)(z-2)} = -\sum_{n=0}^{\infty}(z-1)^n - \frac{1}{z-1}, \quad 0 < |z-1| < 1$$

The singular part contains only the first term, hence $z_0 = 1$ is a simple pole.

Example 45 Find the residue of $f(z) = \frac{e^z}{z-1}$ at $z_0 = 1$.

Solution: We use that the exponential e^z is analytic to Taylor expand it at $z_0 = 1$ as

$$e^z = \sum_{n=0}^{\infty} \frac{1}{n!}\left(\frac{d^n}{dz^n}e^z\right)_{z=1}(z-1)^n = e\sum_{n=0}^{\infty}\frac{1}{n!}(z-1)^n$$

The function is analytic for $|z-1| > 0$ (disk punctuated at $z_0 = 1$ and of arbitrary radius) and can be expanded as

$$\begin{aligned}f(z) &= \frac{e}{z-1}\left(1 + (z-1) + \frac{1}{2!}(z-1)^2 + \frac{1}{3!}(z-1)^3 + \cdots\right) \\ &= \frac{e}{z-1} + e\left(1 + \frac{1}{2!}(z-1) + \frac{1}{3!}(z-1)^2 + \cdots\right) \\ &= \frac{e}{z-1} + e\sum_{n=0}^{\infty}\frac{1}{(n+1)!}(z-1)^n\end{aligned}$$

The singular part has only the first term, hence $z_0 = 1$ is a simple pole and $\text{Res}(1) = e$.

Example 46 Find the residue at $z_0 = 0$ of $f(z) = \frac{1}{z(z+1)}$.

Solution: We use the partial fraction decomposition

$$f(z) = \frac{1}{z(z+1)} = \frac{1}{z} - \frac{1}{z+1}.$$

Since $\frac{1}{z+1}$ is analytic at $z_0 = 0$ and for $|z| < 1$, we can Taylor expand it as the geometric series

$$\frac{1}{z+1} = \sum_{n=0}^{\infty} (-1)^n z^n$$

Hence, for $0 < |z| < 1$, the function is expanded as

$$f(z) = \frac{1}{z(z+1)} = \frac{1}{z} - \sum_{n=0}^{\infty} (-1)^n z^n$$

The singular part has only the first term $b_1 = 1$, hence $z_0 = 1$ is a simple pole and Res(0) = 1.

Essential singularity: If $b_n \neq 0$ for any n, then z_0 is an *essential singularity*.

Behavior of a differentiable function at ∞_:_ Suppose that $f(z)$ is differentiable for $|z| > R$. Then we can define

$$g(w) = f(1/z) \quad 0 < |w| < 1/R.$$

Since $g'(w) = -z^{-2} f'(1/z)$, it follows that $g(w)$ is differentiable for $0 < |w| < 1/R$. By definition, the behavior of $f(z)$ at infinity is the same the behavior $g(w)$ at the origin.

Example 47 Find the Laurent expansion of

$$f(1/z) = e^{1/z}$$

at $z_0 = 0$ for $|z| > 0$ (complex plane punctured at origin).

Solution: We use the fact the Laurent expansion is unique for a given analytic domain. We first consider the simple exponential function $f(w) = e^w$ which is an analytic function everywhere in the complex plane. Its Taylor expansion at $w = 0$ is given by

$$f(w) = e^w = \sum_{n=0}^{\infty} \frac{1}{n!} w^n.$$

This holds also $w = 1/z$,

$$e^{1/z} = f(1/z) = \sum_{n=0}^{\infty} \frac{1}{n!} \frac{1}{z^n}$$

which is a singular series with all $b_n \neq 0$, thus $z_0 = 0$ is an essential singularity.

Example 48 Find the Laurent expansion of

$$f(1/z) = \cos(1/z)$$

at $z_0 = 0$ for $|z| > 0$.

Solution: Again we use the Laurent expansion is unique for a given analytic domain. In this case, we consider $f(w) = \cos(w)$ which is analytic at Taylor expansion at $w = 0$ is given by

$$f(w) = \cos(w) = \sum_{n=0}^{\infty} a_n w^n,$$

where

$$a_n = \frac{1}{n!} \frac{d^n}{dw^n} \cos(w) \Big|_{w=0} \tag{55}$$

$$= \begin{cases} \frac{1}{(2k)!}(-1)^k, & n = 2k \\ 0, & n = 2k+1 \end{cases} \tag{56}$$

Thus,

$$f(w) = \cos(w) = \sum_{k=0}^{\infty} \frac{1}{(2k)!}(-1)^k w^{2k},$$

For $|z| > 0$, the $\cos(1/z)$ is analytic and has a unique expansion. Therefore, it can be obtained by substituting $w = 1/z$ in the Taylor expansion,

$$f(1/z) = \cos(1/z) = \sum_{k=0}^{\infty} \frac{(-1)^k}{(2k)!} \frac{1}{z^{2k}}.$$

8 Lecture 8: Methods of Finding the Residue

8.1 Finding the Residue

The residue of a $f(z)$ at an isolated singular point z_0 captures the "strength" of the singular behavior. Finding the residue of $f(z)$ is useful in evaluating complex integrals such as

$$\oint_{|z-\pi/2|=\pi/2} \tan(z)dz,$$

where isolated singularities are "baked" into the integrand function in such a way that the Cauchy integral formula is not readily applicable. Using the residue theorem, the contour integral of $f(z)$ is determined by the residues at all the isolated singularities enclosed by the contour. We present several methods of finding the residue without actually performing the Laurent expansion.

8.1.1 Laurent Expansion of $f(z)$

The Laurent expansion of $f(z)$ at z_0 for the analytic domain $|z - z_0| > 0$,

$$\boxed{f(z) = \sum_{n=0}^{\infty} a_n(z-z_0)^n + \sum_{n=1}^{\infty} b_n \frac{1}{(z-z_0)^n}} \tag{57}$$

gives us the residue of $f(z)$ at z_0 defined as

$$\boxed{\text{Res}(f, z_0) = b_1} \tag{58}$$

Example 49 Find the residue at $z_0 = \frac{1}{2}$ of $f(z) = \frac{\sin(\pi z)}{4z^2 - 1}$.

Solution: To find the residue at $z_0 = 1/2$, we need to Laurent expand $f(z)$ in the punctured disk $|z - 1/2| > 0$.

The $\sin(\pi z)$ function is analytic at $z_0 = 1/2$. We use the partial fraction decomposition to express the rational function as

$$\frac{1}{4z^2-1} = \frac{1}{(2z-1)(2z+1)} = \frac{1}{2}\left(\frac{1}{2z-1} - \frac{1}{2z+1}\right).$$

$\frac{1}{2z+1}$ is analytic at $z_0 = 1/2$, hence it does not contribute to the residue. Thus, a non-zero residue of $f(z)$ at $z_0 = 1/2$ is due to this function

$$\frac{1}{4}\frac{\sin(\pi z)}{z - 1/2}$$

The $\sin(\pi z)$ has the Taylor expansion at $z_0 = 1/2$ given as:

$$\sin(\pi z) = \sin\left(\frac{\pi}{2}\right) + \pi \cos\left(\frac{\pi}{2}\right)(z - \frac{1}{2}) - \frac{\pi^2}{2!}\sin\left(\frac{\pi}{2}\right)(z - \frac{1}{2})^2 \cdots$$

$$= 1 - \frac{\pi^2}{2!}(z - \frac{1}{2})^2 + \frac{\pi^4}{4!}(z - \frac{1}{2})^4 - \cdots$$

8 Lecture 8: Methods of Finding the Residue

By dividing it with $\frac{1}{4(z-1/2)}$, we find the corresponding Laurent expansions for $|z - 1/2| > 0$:

$$\frac{1}{4}\frac{1}{z-1/2} - \frac{1}{4}\frac{\pi^2}{2!}\left(z - \frac{1}{2}\right) + \frac{1}{4}\frac{\pi^4}{4!}\left(z - \frac{1}{2}\right)^3 - \cdots$$

Thus,
$$\text{Res}(f(z), 1/2) = 1/4.$$

8.1.2 Simple Pole Rule

When z_0 is a simple pole of $f(z)$, then $f(z)$ can be written as

$$f(z) = \frac{g(z)}{z - z_0}$$

with $g(z)$ being analytic at z_0 and $g(z_0) \neq 0$ (z_0 is not a zero of g). The residue at z_0 can be determined by the simple pole rule

$$\boxed{\text{Res}(z_0) = \lim_{z \to z_0} (z - z_0) f(z) = g(z_0)} \tag{59}$$

Example 50 Find the residue at $z_0 = 0$ of $f(z) = \frac{\cos(z)}{z}$.

Solution: First we notice that $zf(z) = g(z) = \cos(z)$ which is analytic at z_0 and $\cos(0) = 1$. Hence, the residue is

$$\text{Res}(0) = \cos(0) = 1$$

Example 51 Find the residue at $z_0 = i$ of $f(z) = \frac{\sin(z)}{(1-z^4)}$.

Solution:
$$f(z) = \frac{\sin(z)}{(z-i)(z+i)(1-z^2)} = \frac{g(z)}{z-i},$$

where
$$g(z) = \frac{\sin(z)}{(z+i)(1-z^2)}$$

which is analytic at $z_0 = i$ and $g(i) = \sin(i)/(4i)$. Hence $z_0 = i$ is a simple pole,

$$f(z) = \frac{g(i)}{z-i} + \text{regular part}$$

Hence, the residue is

$$\text{Res}(i) = g(i) = \frac{\sin(i)}{4i} = \frac{\sinh(1)}{4}$$

Theorem 9 (L'Hôpital rule for a simple pole) *When*

$$\boxed{f(z) = \frac{g(z)}{h(z)}}$$

with $g(z)$ and $h(z)$ being analytic at z_0 and

$$g(z_0) \neq 0, \quad h(z_0) = 0, \quad h'(z_0) \neq 0,$$

then $f(z)$ has a simple pole at z_0 with the residue determined by the L'Hôpital rule:

$$\boxed{\text{Res}(z_0) = \frac{g(z_0)}{h'(z_0)}} \tag{60}$$

Proof

$$\begin{aligned}
\text{Res}(z_0) &= \lim_{z \to z_0} \frac{(z - z_0)g(z)}{h(z)} \\
&= g(z_0) \lim_{z \to z_0} \frac{(z - z_0)}{h(z)} \\
&= g(z_0) \lim_{z \to z_0} \frac{1}{h'(z)} \\
&= \frac{g(z_0)}{h'(z_0)}
\end{aligned}$$

Alternative way to determine that $f(z)$ has a simple pole at z_0: By comparing the smallest powers in the Taylor expansions of $g(z)$ and $h(z)$ at z_0. When $h(z)$ has the leading order term **one** order higher than the leading order term in $g(z)$, then z_0 is a simple pole.

Example 52 Find the residue at $z_0 = 0$ of $f(z) = \frac{\cos(2z)}{\sin(2z)}$.

Solution: The cos and sin functions are analytic. Furthermore,

$$\cos(0) = 1, \quad \sin(0) = 0, \quad \frac{d\sin(2z)}{dz}\bigg|_{z=0} = 2\cos(0) = 2.$$

Hence, z_0 is a simple pole and the residue is

8 Lecture 8: Methods of Finding the Residue

$$\text{Res}(0) = \frac{\cos(0)}{\cos(0)} = \frac{1}{2}.$$

Alternatively, we could look at the first terms in the Taylor expansion of $\sin(2z)$

$$\sin(2z) = \sin(0) + 2\cos(0)z - 4\sin(0)\frac{z^2}{2} - 8\cos(0)\frac{z^3}{3!} + \mathcal{O}(z^4)$$
$$= 2z - 4\frac{z^3}{3} + \mathcal{O}(z^4)$$

and compare it with that of $\cos(2z)$

$$\cos(2z) = \cos(0) - 2\sin(0)z - 4\cos(0)\frac{z^2}{2} + \mathcal{O}(z^3)$$
$$= 1 - 2z^2 + \mathcal{O}(z^3)$$

Hence, by dividing them we find that

$$\frac{\cos(2z)}{\sin(2z)} = \frac{1 - 2z^2 + \mathcal{O}(z^3)}{2z - 4\frac{z^3}{3} + \mathcal{O}(z^4)}$$
$$= \frac{1}{2z}\frac{1}{1 - 2\frac{z^2}{3} + \mathcal{O}(z^3)} - \frac{z + \mathcal{O}(z^2)}{1 - 2\frac{z^2}{3} + \mathcal{O}(z^3)}$$

from which we may see that in the limit where z approaches 0, the leading order singular term is

$$\frac{1}{2z}$$

and thus the residue is $1/2$.

8.1.3 Multiple Pole Rule

If z_0 is a pole of order n of $f(z)$, then the residue at z_0 is

$$\boxed{\text{Res}(z_0) = \frac{1}{(m-1)!}\frac{d^{m-1}}{dz^{m-1}}[(z-z_0)^m f(z)]\Big|_{z=z_0}, \quad m \geq n} \qquad (61)$$

Proof Let us use the Laurent expansion for $f(z)$ at z_0 which is a pole of order n. This means that the singular part is a finite sum with n terms:

$$\frac{b_1}{z - z_0} + \cdots + \frac{b_n}{(z - z_0)^n} + \sum_{k=0}^{\infty} a_k(z - z_0)^k.$$

By multiplying with the power term $(z - z_0)^m$ with $m \geq n$, we can eliminate all the singular terms

$$(z - z_0)^m f(z) = b_1(z - z_0)^{m-1} + b_2(z - z_0)^{m-2} + \cdots + b_n(z - z_0)^{m-n}$$
$$+ \sum_{k=0}^{\infty} a_k (z - z_0)^{k+m}$$

By taking the $(m - 1)$ derivative of this and evaluate it at z_0, we can determine an expression for the b_1 coefficient

$$\frac{d^{m-1}}{dz^{m-1}}[(z - z_0)^m f(z)]\bigg|_{z=z_0} = (m - 1)! b_1$$

and thus determine the residue

$$\text{Res}(z_0) = b_1 = \frac{1}{(m - 1)!} \frac{d^{m-1}}{dz^{m-1}}[(z - z_0)^m f(z)]\bigg|_{z=z_0}$$

Example 53 Find the residue at $z_0 = \pi$ of $f(z) = \frac{z \sin z}{(z-\pi)^3}$.

Solution: The nominator $z \sin(z)$ is analytic, hence $z_0 = \pi$ is a pole of order of at most 3. So, we can use $m = 3$ and apply the multiple pole rule

$$\text{Res}(\pi) = \frac{1}{2!} \frac{d^2}{dz^2}[z \sin z]\bigg|_{z=\pi}$$
$$= \frac{1}{2} \frac{d}{dz}[\sin z + z \cos z]\bigg|_{z=\pi}$$
$$= \frac{1}{2}[2 \cos z - z \sin z]\bigg|_{z=\pi}$$
$$= -1$$

Example 54 Find the residue at $z_0 = 0$ of

$$f(z) = \frac{\cos(z)}{z^4}$$

Solution: $\cos(z)$ function is analytic at $z_0 = 0$ and has the Taylor expansion

$$\cos(z) = 1 - \frac{1}{2}z^2 + \frac{1}{4!}z^4 \cdots$$

Thus,

$$f(z) = \frac{\cos(z)}{z^4} = \frac{1}{z^4} - \frac{1}{2z^2} + \text{regular terms}$$

We notice something very interesting here: the singular part has the highest order negative power given by $n = 4$ which means that we are dealing with pole of order 4, but due to the symmetry of the function, the first term vanishes and therefore $b_1 = 0$ which implies that $Res(0) = 0$ (as if this were a regular point). So, what is happening here? This a great example of an isolated singularity which does not contribute to the contour integral, because the simple fractions

$$\frac{1}{z^2}, \ \frac{1}{z^4},$$

which, even though clearly singular at $z_0 = 0$, give no contribution to the contour integrals, i.e.

$$\oint_{|z|=R} \frac{dz}{z^2} = 0$$

and

$$\oint_{|z|=R} \frac{dz}{z^4} = 0$$

by virtue of the generalised Cauchy's integral formula.

8.2 Residue Theorem

The residue theorem is very useful in evaluating complex integrals as well as many definite real integrals.

Theorem 10 *Let z_0 be an isolated singular point of $f(z)$ and C be a positive-oriented simple contour **enclosing z_0 and no other singularities**. Then, the integral of $f(z)$ on C is determined by the residue of $f(z)$ at z_0:*

$$\boxed{\oint_C f(z)dz = 2\pi i Res(f, z_0),} \qquad (62)$$

with $Res(z_0) = b_1$ is the residue of $f(z)$ at z_0 determined by the Laurent expansion of $f(z)$ at z_0 for $|z - z_0| > 0$ and enclosed by the contour C.

Proof For any $z \neq z_0$ inside C, $f(z)$ is analytic, hence it has a well-defined Laurent expansion. Now, the regular part of the Laurent series is analytic everywhere including at z_0. Hence by the Cauchy's theorem, its contour integral vanishes term by term, namely

$$\sum_{n=0}^{\infty} a_n \oint_C (z - z_0)^n dz = 0.$$

This means that the contour integral of $f(z)$ is non-zero because it picks up a net contribution from the singular part of the Laurent series:

$$\oint_C f(z) dz = \sum_{n=1}^{\infty} b_n \oint_C \frac{1}{(z - z_0)^n} dz \tag{63}$$

It turns out that only the first term survives this integral while all the higher order terms vanish upon integration. From the Cauchy's integral formula, the contour integral of the first term reduces to the b_1 coefficient

$$b_1 \oint_C \frac{1}{(z - z_0)} dz = 2\pi i b_1,$$

and therefore it is the *residue*, as the only non-zero contribution to the contour integral. All the higher order terms, vanish by the generalized Cauchy's formula:

$$\oint_C \frac{dz}{(z - z_0)^{n+1}} = 0$$

Hence, the contour integral of $f(z)$ is determined by its residue at z_0 as (Fig. 18)

$$\oint_C f(z) dz = 2\pi i b_1$$

Theorem 11 (Residue theorem II) *Let $z_0, z_1, z_2, \ldots$ be isolated singularities of $f(z)$. Then, the integral of $f(z)$ around a simple closed curve C and surrounding all singular points is determined by:*

$$\boxed{\oint_C f(z) dz = 2\pi i \sum_k Res(z_k).} \tag{64}$$

Proof We draw a small circle around each isolated singularity and connect them through cuts with the main closed curve. By Cauchy's theorem:

Fig. 18 Analytic domain around isolated singularities

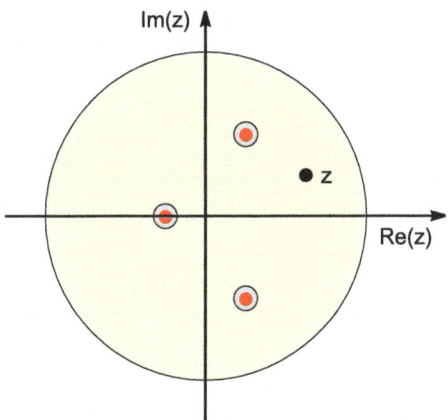

$$\oint_{C-C_0-C_1-C_2-\cdots} f(z)dz = 0$$

which implies that the integral over the main contour is determined by the sum of the integrals over the inner, small circles enclosing the singularities

$$\oint_C f(z)dz = \sum_k \oint_{C_k} f(z)dz.$$

Since inside each circle we have only one isolated singularity z_i, the contour integral with z_k inside is given by the b_1 of the Laurent series expansion of $f(z)$ at z_k, i.e. the residue of $f(z)$ at z_k, Res(z_k). Hence,

$$\oint_C f(z)dz = 2\pi i \sum_k \text{Res}(z_k)$$

Example 55 Use the residue theorem to evaluate this integral (Fig. 19)

$$I = \oint_{|z|=4} \frac{z+2}{z^2+9} dz.$$

Solution: Let us rewrite the integrand function as

$$f(z) = \frac{z+2}{z^2+9} = \frac{z+2}{(z+3i)(z-3i)}$$

Fig. 19 Analytic domain for Example 55

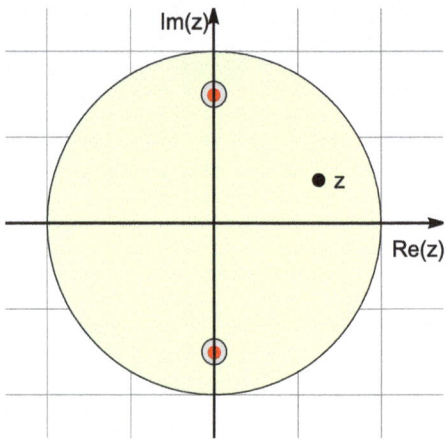

such that we can identify two simple poles at $z_0 = \pm 3i$. Both points are inside the disk $|z| < 4$. Hence, the integral is determined by the residue of $f(z)$ at these poles.

$$\oint_{|z|=4} \frac{z+2}{z^2+9} dz = 2\pi i [\text{Res}(3i) + \text{Res}(-3i)]$$

The residues can be quickly determined by the simple pole rule:

$$\text{Res}(3i) = \lim_{z \to 3i} (z - 3i) f(z) = \frac{3i+2}{6i} = \frac{1}{2} - i\frac{1}{3}$$

$$\text{Res}(-3i) = \lim_{z \to 3i} (z + 3i) f(z) = \frac{-3i+2}{-6i} = \frac{1}{2} + i\frac{1}{3}$$

Hence, $I = 2\pi i$.

Example 56 Evaluate the integral (Fig. 20)

$$\oint_{|z-\pi/2|=\pi/2} \tan(z) dz,$$

Solution: We notice that $\tan(z) = \frac{\sin z}{\cos z}$ is analytic inside the disk $|z - \pi/2| \leq \pi/2$ except at $z_0 = \pi/2$. Thus, the residue theorem,

$$\oint_{|z-\pi/2|=\pi/2} \tan(z) dz = 2\pi i \, \text{Res}\left(\tan(z), z = \frac{\pi}{2}\right)$$

Fig. 20 Analytic domain for Example 56

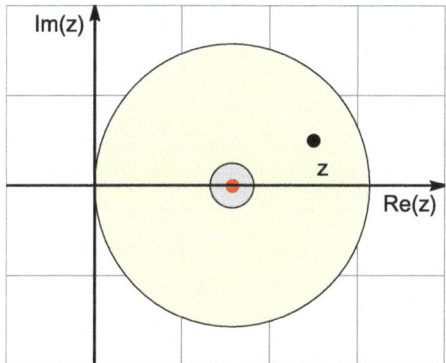

To determine the order of this pole, we Taylor expand to lowers order the trigonometric functions as

$$\sin(z) = 1 - \frac{1}{2}\left(z - \frac{\pi}{2}\right)^2 + \cdots,$$

$$\cos(z) = -\left(z - \frac{\pi}{2}\right) + \frac{1}{3!}\left(z - \frac{\pi}{2}\right)^3 + \cdots,$$

Thus, $z = \pi/2$ is a simple pole and we apply the L'Hôpital rule to determine the residue

$$Res\left(\tan(z), z = \frac{\pi}{2}\right) = \frac{\sin(\pi/2)}{-\sin(\pi/2)} = -1.$$

9 Lecture 9: Definite Integrals

Complex analysis is a powerful toolbox for evaluating definite integrals that can be more difficult to handle within real analysis. In this lecture, we introduce various techniques for evaluating different types of definite integrals depending on the types of integrand functions.

9.1 Mapping to Unit Circle

For an integral of this form

$$I = \int_0^{2\pi} f(\theta)d\theta$$

it is useful to change the integration variable as $z = e^{i\theta}$, such the integral transforms into a contour integral over the unit circle $|z| = 1$. Namely,

$$z = e^{i\theta} \to dz = ie^{i\theta}d\theta \to d\theta = \frac{dz}{iz},$$

and

$$\int_0^{2\pi} f(\theta)d\theta = -i \oint_{|z|=1} \frac{f(z)}{z}dz \tag{65}$$

which can be evaluated by the residue theorem for $f(z)/z$ when none of it poles are on the circle.

Example 57 Evaluate the definite integral

$$I = \int_0^{2\pi} \frac{d\theta}{2 + \sin(\theta)}.$$

Solution:

We use the complex representation of $\sin(\theta)$ follows from the Euler's formula:

$$\sin(\theta) = \frac{e^{i\theta} - e^{-i\theta}}{2i} = \frac{z - z^{-1}}{2i}$$

Hence, the integral mapped to the contour integral is

$$I = \oint_{|z|=1} \frac{dz}{iz} \frac{2i}{4i + (z - 1/z)}$$

$$= \oint_{|z|=1} \frac{dz}{iz} \frac{2iz}{z^2 + 4iz - 1}$$

$$= \oint_{|z|=1} dz \frac{2}{(z + i(2 + \sqrt{3}))(z + i(2 - \sqrt{3}))}.$$

We recognise that the integrand has two simple poles at $z_0 = i(2 \pm \sqrt{3})$, but only one of them is inside the unit circle, as shown in Fig. 21. Thus, by the residue theorem, we can evaluate this integral in terms of the residue at $z_0 = -i(2 - \sqrt{3})$

$$I = 2\pi i \operatorname{Res}\left(z_0 = -i(2 - \sqrt{3})\right),$$

where the residue can be quickly evaluated from the simple pole rule

$$\operatorname{Res}\left(-i(2 - \sqrt{3})\right) = \lim_{z \to z_0} \frac{2}{(z + i(2 + \sqrt{3}))} = \frac{2}{2i\sqrt{3}}.$$

9 Lecture 9: Definite Integrals

Fig. 21 Illustration for Example 57

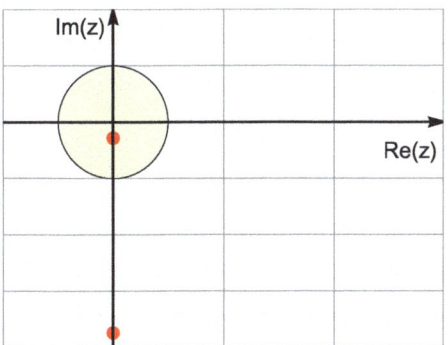

Thus, the definite integral equals to

$$I = \frac{2\pi}{\sqrt{3}}.$$

9.2 Extend to Complex Plane

Let us consider an integral of this generic form

$$I = \int_{-\infty}^{\infty} f(x)dx,$$

where the integrand function satisfies the following conditions when extended to the complex plane:

1. $f(z)$ is analytic in the complex plane except for a finite number of poles, none on the real axis
2. $f(z)$ is upper-bounded on the semi-circle Γ_ρ in one of the half-planes as

$$|f(z)| \leq \frac{A}{\rho^2}, \text{ for } |z| = \rho \text{ and } A \text{ a constant}$$

Then, this integral is equal to the contour integral over one the half-planes

$$\boxed{\int_{-\infty}^{+\infty} f(x)dx = \oint_C f(z)dz} \tag{66}$$

which can be evaluated by the residue theorem. In most examples this applies for a rational integrand

$$f(x) = \frac{P(x)}{Q(x)}$$

where $P(x)$ and $Q(x)$ are polynomials, such that $deg(P) \le deg(Q) - 2$ and $Q(x)$ has no zeros on the real axis.

Example 58 Evaluate the integral

$$I = \int_{-\infty}^{\infty} \frac{dx}{1+x^2}$$

Solution: Let us examine the contour integral in the upper half-plane:

$$I_C = \oint_C \frac{dz}{1+z^2} = \oint_C \frac{dz}{(z+i)(z-i)},$$

where C is the closed loop with the upper semicircle Γ_ρ as shown in Fig. 22. The integrand has two simple poles but only $z_0 = i$ is enclosed by C. The residue of the integrand at this point follows from the simple pole rule as

$$\text{Res}(i) = \frac{1}{(z+i)}\bigg|_{z=0} = \frac{1}{2i}.$$

Hence, by residue theorem

$$I_C = 2\pi i \text{Res}(i) = \pi.$$

Now, we want to relate I_C to the real integral I. For this, we decompose the contour into the integral over the real axis and the integral over the semicircle Γ_ρ:

$$I_C = \pi = \int_{-\rho}^{\rho} \frac{dx}{1+x^2} + \int_{\Gamma_\rho} \frac{dz}{1+z^2}$$

Fig. 22 Illustration for Example 58

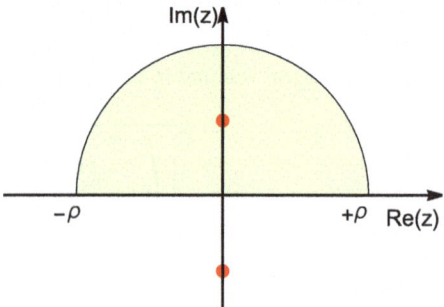

such that the integral can be expressed as

$$I = \int_{-\infty}^{\infty} \frac{dx}{1+x^2} = \pi - \lim_{\rho \to \infty} \int_{\Gamma_\rho} \frac{dz}{1+z^2}.$$

We use the Cauchy's inequality to show that the line integral on the right hand side vanishes in the limit of infinite ρ. The integrand $f(z) = 1/(1+z^2)$ is upper bounded on Γ_ρ by:

$$|f(z)| \leq \frac{1}{|1+z^2|} \leq \frac{1}{|z|^2 - 1} = \frac{1}{\rho^2 - 1}$$

Hence, by the Cauchy's inequality, the magnitude of the integral is upper-bounded by

$$\left| \int_{\Gamma_\rho} f(z) dz \right| \leq \pi \rho |f(z)| \leq \frac{\pi \rho}{\rho^2 - 1},$$

which vanishes in the limit $\rho \to \infty$, i.e.

$$\lim_{\rho \to \infty} \left| \int_{\Gamma_\rho} f(z) dz \right| \leq \lim_{\rho \to \infty} \frac{\pi \rho}{\rho^2 - 1} = 0$$

Hence, $I = \pi$. Often, it suffices to check that the degree of polynomial in the denominator is higher then that of the nominator. In this example, $deg(Q) = 2$ and $deg(P) = 0$.

9.3 Jordan's Lemma

Let us consider an integral of this generic form

$$I = \int_{-\infty}^{\infty} f(x) e^{ix} dx,$$

where $f(z)$ satisfies the following conditions in the complex plane:

1. $f(z)$ is analytic in the upper-half plane except for a finite number of poles, none on the real axis.
2. $f(z)$ is upped-bounded on the semicircle Γ_ρ as

$$|f(z)| \leq \frac{A}{\rho} \quad \text{for } |z| = \rho \text{ and A is a constant.}$$

Then, the integral can be extended to the *upper-half plane* and equals the contour integral

$$\int_{-\infty}^{+\infty} f(x)e^{ix}dx = \oint_C f(z)e^{iz}dz. \tag{67}$$

This can be generalized to any integer power m of the complex exponential

$$\int_{-\infty}^{+\infty} f(x)e^{imx}dx = \oint_C f(z)e^{imz}dz, \tag{68}$$

where the contour C is in the *upper-half* plane for $m > 0$ and C is in the *lower-half* plane for $m < 0$. The half-plane is selected such that the corresponding line integral over the semi-circle vanishes in the limits of infinite radius. This is related to the asymptotic behavior of the complex exponential in the plane

$$|e^{imz}| = |e^{imx}||e^{-my}| = e^{-my},$$

which vanish in the upper half-plane for $m > 0$, and in the lower half-plane for $m < 0$.

Example 59 Evaluate the following integral:

$$I = \int_{-\infty}^{+\infty} dx \frac{x \sin x}{1+x^2}.$$

Solution: First, we rewrite this as:

$$I = \frac{1}{2i} \left[\int_{-\infty}^{+\infty} \frac{xe^{ix}}{1+x^2} dx - \int_{-\infty}^{+\infty} \frac{xe^{-ix}}{1+x^2} dx \right] = \text{Im}\left(\int_{-\infty}^{+\infty} \frac{x}{1+x^2} e^{ix} dx \right)$$

Using that

$$|e^{iz}| = e^{-y}$$

we select the upper-half plane $y > 0$, where the exponential is decaying. Let us now evaluate the corresponding contour integral in this *upper-half* plane

$$I_{C_p} = \oint_{C_p} \frac{ze^{iz}}{(z+i)(z-i)} dz$$

$$= 2\pi i \, \text{Res}\left(\frac{ze^{iz}}{(z+i)(z-i)}, i \right)$$

$$= \frac{i\pi}{e},$$

9 Lecture 9: Definite Integrals

where C_ρ contains the semicircle on the positive half-plane. The integral over the semi-circle vanishes as its upper bound vanishes in the limit of infinite radius by virtue of the Cauchy's inequality:

$$\left| \int_{\Gamma_\rho} \frac{ze^{iz}}{z^2+1} dz \right| \leq \frac{\pi \rho^2 e^{-\rho}}{\rho^2 - 1} \xrightarrow[\rho \to \infty]{} 0.$$

Hence,

$$I = \mathrm{Im}\left(\frac{i\pi}{e}\right) = \frac{\pi}{e}$$

9.4 Improper Integrals

9.4.1 Cauchy's Principal Value

The principal value method applies to improper integrals that have a singular behavior around isolated points. For an integral

$$\int_{-\infty}^{\infty} f(x) dx$$

which has a singularity at x_0, we can extract the non-singular value by *the principal value (P.V.)* integral defined as

$$\mathrm{P.V.} \int_{-\infty}^{\infty} f(x) dx = \lim_{\epsilon \to 0} \left[\int_{-\infty}^{x_0 - \epsilon} f(x) dx + \int_{x_0 + \epsilon}^{\infty} f(x) dx \right].$$

This principal value integral can be determined by complex integration. The idea is to introduce the contour integral in the half-plane as illustrated in Fig. 23 to isolate the

Fig. 23 Contour circumventing simple poles on the real axis

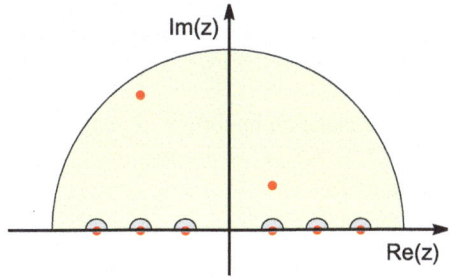

singularities located at x_j on the real axis. The contour integral can be decomposed into

$$\oint_C f(z)dz = P.V. \int_{-\infty}^{+\infty} f(x)dx - \lim_{\epsilon \to 0} \sum_j \int_{\Gamma_{\epsilon,j}} f(z)dz + \int_{\Gamma_\rho} f(z)dz,$$

where the minus sign on the lhs accounts for the fact the small semi-circles $\Gamma_{\epsilon,j}$ (centered at the poles x_j and of radius ϵ) have a negative orientation (clockwise). Now, the line integral over the large semi-circle Γ_ρ vanishes in the limit of infinite radius since $f(z)$ decays sufficiently fast (the same argument as in the other classes of integrand functions). This works when the integrand is of the form $f(z) = \frac{P(z)}{Q(z)}$ where $deg(Q) = deg(P) + 2$ or $f(z) = \frac{P(z)}{Q(z)} e^{imz}$. In the latter case, we choose the half-plane for which the complex exponential decays with y (Jordan's lemma). Then, we have that

$$\left| \int_{\Gamma_\rho} f(z)dz \right| \to 0 \text{ as } \rho \to \infty$$

Thus,

$$\boxed{P.V. \int_{-\infty}^{+\infty} f(x)dx = \oint_C f(z)dz + \lim_{\epsilon \to 0} \sum_j \int_{\Gamma_{\epsilon,j}} f(z)dz.} \qquad (69)$$

We can evaluate the contour integral over C by the residue theorem when $f(z)$ has isolated singularities,

$$\oint_C f(z)dz = 2\pi i \sum_k \text{Res}(f(z), z_k).$$

When the isolated singularities on the real axis are **simple** poles x_j then we can also evaluate the line integrals over the small semi-circles in terms of the residues at x_j. To show this, we use the line parameterization on the small semicircle centered at x_j given by $z = x_j + \epsilon e^{i\theta}$, such the

$$\lim_{\epsilon \to 0} \int_{\Gamma_{\epsilon,j}} f(z)dz = \lim_{\epsilon \to 0} \int_0^\pi f(x_j + \epsilon e^{i\theta}) i\epsilon e^{i\theta} d\theta$$

Now, we make an important assumption about the structure of $f(z)$ near the poles x_j. It is here where we use that the x_j are **simple** poles to write the function on this semi-circle as

$$f(z) = \frac{g(z)}{z - x_j}$$

where $g(z)$ is analytic at x_j. Inserting this into the line integral from above, we have that

$$\lim_{\epsilon \to 0} \int_{\Gamma_{\epsilon,j}} f(z)dz = \lim_{\epsilon \to 0} \int_0^\pi \frac{g(x_j + \epsilon e^{i\theta})}{\epsilon e^{i\theta}} i\epsilon e^{i\theta} d\theta$$

$$= ig(x_j) \int_0^\pi d\theta$$

$$= i\pi g(x_j) = i\pi \text{Res}(f(z), x_j).$$

We have used that $g(z)$ is analytic at x_j thus the limit of zero ϵ is well-defined and equal to the value $g(x_j)$ which also determines the residue of $f(z)$ at x_j (the simple pole rule). Notice that each pole on the real axis contributes with a **factor** πi (instead of $2\pi i$) corresponding to being enclosed by a semi-circle!

Finally, collecting all these residues, we can express the principal value of the integral as

$$\boxed{P.V. \int_{-\infty}^{+\infty} f(x)dx = 2\pi i \sum_k \text{Res}(f(z), z_k) + \pi i \sum_j \text{Res}(f(z), x_j)} \qquad (70)$$

Let us take two examples to illustrate how to apply the Cauchy principal value method in practice. One in which the integrand is on the form $f(z) = \frac{P(z)}{Q}$ and the other on the form $f(z) = \frac{P(z)}{Q} e^{imz}$ (Fig. 24).

Example 60 Evaluate this integral

$$I = P.V. \int_{-\infty}^{+\infty} \frac{1}{(x+1)(x^2+4)} dx$$

Solution: We notice that the integrand as a simple pole on the real axis $x_0 = -1$. Additionally, there are two simple poles on the imaginary axis at $z = \pm 2i$. The

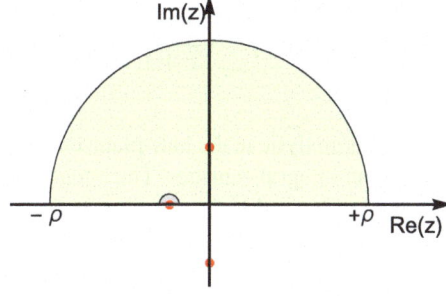

Fig. 24 Contour circumventing the singular point at $z_0 = -1$ from Example 60

complex integral over the outer half-circle Γ_ρ vanishes in the limit of an infinite radius since the integrand is rational

$$\frac{P(x)}{Q(x)} = \frac{1}{(x+1)(x^2+4)},$$

and that $\deg(P) = 0$, while $\deg(Q) = 3$ (same argument as for integrals with rational integrands).

Thus, the principal value is determined by the contour integral in the half-plane and the semicircle of radius ϵ centered at the pole.

$$I = \oint_C \frac{1}{(z+1)(z^2+4)} dz + \lim_{\epsilon \to 0} \int_{\Gamma_{\epsilon,-1}} \frac{1}{(z+1)(z^2+4)} dz,$$

where $\Gamma_{\epsilon,-1}$ is the semi-circle of radius ϵ centered at -1. Using the residue at simple pole -1 and $2i$ in the upper-half plane, we then evaluate the principal value as

$$I = 2\pi i \operatorname{Res}(2i) + \pi i \operatorname{Res}(-1) = \frac{\pi(1-2i)}{10} + i\frac{\pi}{5} = \frac{\pi}{10}.$$

Example 61 Evaluate this integral

$$I = P.V. \int_{-\infty}^{+\infty} \frac{\sin(\pi x)}{x+1} dx$$

Solution: Let us rewrite the integral equivalently in terms of the complex exponential

$$I = P.V. \int_{-\infty}^{+\infty} \frac{\sin(\pi x)}{x+1} dx = \operatorname{Im}\left[P.V. \int_{-\infty}^{+\infty} \frac{e^{i\pi x}}{x+1} dx \right]$$

By Jordan's lemma, the function $f(z) = \frac{e^{i\pi z}}{z+1}$ is decaying in the upper half-plane $y > 0$, thus we take the contour C_+ in this domain. By the Cauchy principal value method, the integral reduces to

$$P.V. \int_{-\infty}^{+\infty} \frac{e^{i\pi x}}{x+1} dx = \oint_{C_+} \frac{e^{i\pi z}}{z+1} dz + \lim_{\epsilon \to 0} \int_{\Gamma_{\epsilon,-1}} \frac{e^{i\pi z}}{z+1} dz.$$

The $f(z)$ is analytic in the half-plane enclosed by C_+, thus by the Cauchy's theorem, this contour integral vanishes. The integral over the semi-circle centered at the simple pole at $x_0 = -1$ reduces

$$P.V. \int_{-\infty}^{+\infty} \frac{e^{i\pi x}}{x+1} dx = i\pi \operatorname{Res}\left(\frac{e^{i\pi z}}{z+1}, -1\right)$$

9 Lecture 9: Definite Integrals

$$= i\pi e^{-i\pi} = i\pi\left[\cos(\pi) - i\sin(\pi)\right]$$
$$= -i\pi.$$

Thus, the solution to the integral is $I = -\pi$.

9.4.2 Improper Integral with Branch Cuts (Optional)

We consider one example of an improper integral with a branch cut to illustrate the technique of integrating around a keyhole loop.

Example 62 Evaluate the principal value of this integral

$$I = P.V. \int_0^\infty \frac{1}{(x+1)\sqrt{x}} dx$$

Solution: The function $f(z) = \sqrt{z}$ is multi-valued. To see this, we use the polar representation $z = re^{i\theta}$, with $\theta \in [0, 2\pi]$, such that $f(z) = e^{\frac{1}{2}\ln z} = \sqrt{r}e^{i\theta/2}$. For any θ, if we apply a 2π-rotation and evaluate the function, we see that

$$\sqrt{z} = \begin{cases} \sqrt{r}e^{i\theta/2} \\ \sqrt{r}e^{i(\theta+2\pi)/2} = -\sqrt{r}e^{i\theta/2}. \end{cases}$$

Thus, the function has a branch point at $z = 0$. We choose which branch cut we want to work with by defining the principal value of θ. For our purpose, we place the branch cut on the non-negative real axis, $x \geq 0$. This implies that $\theta \in (0, 2\pi)$. For $\theta \in (-\pi, \pi)$, the branch cut is on the non-positive real axis $x \leq 0$.

To avoid crossing the branch-cut, we construct the contour C as a keyhole loop (see Fig. 25), so that we can evaluate the contour integral over C using residue theorem,

$$I_C = \oint_C \frac{1}{\sqrt{z}(z+1)} dz = 2\pi i \operatorname{Res}(-1) = \frac{2\pi i}{\sqrt{-1}} = 2\pi.$$

We decompose the contour integral as a sum of the integrals over individual segments of the contour, namely

$$I_C = \left[\int_{\Gamma_\rho} + \int_{\Gamma_\epsilon} + \int_{AB} + \int_{DE}\right] \frac{1}{\sqrt{z}(z+1)} dz$$

The integrals over Γ_ρ and Γ_ϵ vanish. We can see this by the transformation of coordinates

$$z = re^{i\theta} \Rightarrow dz = ie^{i\theta} d\theta,$$

Fig. 25 Key-hole contour for Example 62

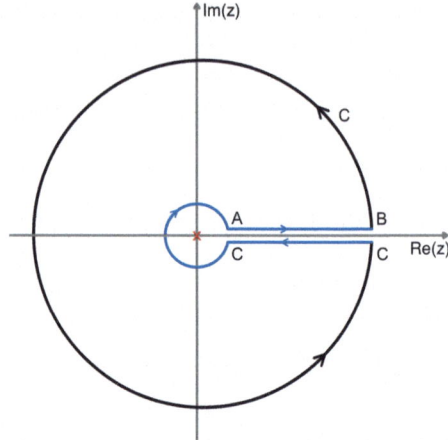

such that

$$\int_{\Gamma_{\rho,\epsilon}} \frac{1}{\sqrt{z}(z+1)} dz = i \int_{\Gamma_{\rho,\epsilon}} \frac{\sqrt{r}e^{i\theta/2} d\theta}{1+re^{i\theta}} \to 0$$

since the integrand vanishes both in the limit of $r \to 0$ and $r \to \infty$.

Thus, we are left with the integrals above and below the branch cut.

On AB: we set $\theta = 0$, hence $z = r$:

$$\int_0^\infty \frac{1}{\sqrt{r}(r+1)} dr = I$$

On DE: we set $\theta = 2\pi$, hence $z = re^{2\pi i}$:

$$\int_{-\infty}^0 \frac{e^{-i\pi}}{\sqrt{r}(r+1)} dr = \int_0^\infty \frac{1}{\sqrt{r}(r+1)} dr = I$$

The keyhole contour is equal to the sum of these integrals, hence

$$I_C = 2I \Rightarrow I = \frac{I_C}{2} = \pi.$$

Variational Calculus

1 Lecture 10: Calculus of Variations: Euler Stationarity Condition

In many physics problems, we may need to determine equilibrium states, find optimal geometrical shapes, or determine evolution equations. *Variational calculus* or *calculus of variations* is a powerful tool to solve these kind of problems.

In variational calculus, the method of finding stationary points—like minima, maxima, or inflection points—is extended to finding stationary curves, surfaces or other complex entities. In physics, the equilibrium state of a system often corresponds to the minimum of a global quantity, such as total energy. For a given function $f(x)$, we find its extreme points by setting its first derivative to zero, $f'(x) = 0$. Variational calculus generalizes this stationarity condition to a **functional** $I[y_1, y_2, y_3 \ldots]$.

A real function takes a real number as input and produces another real number as output, which we write as $f : \mathcal{R} \to \mathcal{R}$. A functional, on the other hand, is a mapping from the space of function to real numbers. We denote this as

$$I : \mathcal{R}^\infty \to \mathcal{R}$$

The functional I is typically defined as a definite integral. In its general form, it looks like this:

$$I[y_1, y_2, y_3 \ldots] = \int_{x_1}^{x_2} F(x, y_1(x), y_2(x), y_3(x) \ldots) dx$$

where the integrand F can depend explicitly on the variable x, or implicitly through other functions $y_1 = y_1(x)$, $y_2 = y_2(x)$, $y_3 = y_3(x)$, etc., which are independent.

In this course, we focus on a specific class of functionals that depend on a real function of one variable $y(x)$, and its first derivative $y' = \frac{dy}{dx}$. The functional I is defined by this definite integral

$$I[y, y'] = \int_{x_1}^{x_2} F(x, y, y')dx.$$

In today's lecture, we will learn how to derive the stationarity condition, commonly known as the Euler condition, using calculus of variations.

1.1 Euler Condition

Our goal is to find the function $y(x)$ that makes the functional $I[y, y']$ stationary. In practice, this optimal function $y(x)$ could represent various entities:

- Geodesic: This is the shortest path between two points on a surface. The functional we want to minimize here is the total distance between these points, and the stationarity condition gives us the equation that defines the geodesic.
- Path of Least Time (Fermat's principle): In this case, the optimal function describes the path that light takes to travel between two points in the least amount of time. The functional represents the total time spent, and minimizing it gives us the path of least time.
- Trajectory of Stationary Action (Hamilton's principle): Here, the function $y(x)$ corresponds to the trajectory of a system. By minimizing the action, which is the functional in this context, we find the equation of motion for the system.

The condition for stationarity includes not only minima but also maxima and inflection points. In many physics problems, the functional (often related to an energy quantity) is minimized to find the equilibrium state, which corresponds to a minimum. However, in optimization problems, variational calculus can be used to find stationary solutions, not just the minima.

1.1.1 Calculus of Variations

Let us apply the calculus of variations to this type of functional

$$I[y, y'] = \int_{x_1}^{x_2} F(x, y(x), y'(x))dx, \tag{1}$$

that depends on $y = y(x)$ and its first derivative $y'(x)$.

Our aim is to determine the optimal curve $y_0(x)$ for which $I[y, y']$ is stationary. This means that **any** variations (infinitesimal perturbations) around the stationary curve $y_0(x)$ will leave $I[y, y']$ unchanged.

Tuning Parameter Method: For this let us consider an **arbitrary** curve $\eta(x)$ with the boundary conditions $\eta(x_1) = \eta(x_2) = 0$ (we will see why, later) and which is superimposed to the stationary one:

1 Lecture 10: Calculus of Variations: Euler Stationarity Condition

$$y(x) = y_0(x) + \epsilon \eta(x),$$

where ϵ is a free tuning parameter such that when $\epsilon \to 0$, we obtain the desired stationary curve. This procedure means that we consider a variational of the curve, and often the variation is denoted as (see below the derivation using this variational notation)

$$\delta y \equiv \epsilon \eta(x).$$

Using that the functional remains unchained by this *arbitrary variation*, we can write it as

$$I[y_0, y_0'] = I[y, y'] = I[y_0 + \epsilon \eta, y_0' + \epsilon \eta'] \equiv I(\epsilon),$$

which means that I as a regular function of the tuning parameter ϵ. This is nice, because now we can use the stationarity condition applied to an ordinary function, namely that its first derivative evaluated at the stationary point must vanish, i.e.

$$\frac{dI}{d\epsilon}\bigg|_{\epsilon=0} = 0. \tag{2}$$

Inserting the integral expression from Eq. 1 into Eq. 2, we obtain that

$$\frac{dI}{d\epsilon} = \int_{x_1}^{x_2} \frac{d}{d\epsilon} F(x, y_0(x) + \epsilon \eta(x), y_0'(x) + \epsilon \eta'(x)) dx$$

$$= \int_{x_1}^{x_2} \left[\frac{\partial F}{\partial y} \eta(x) + \frac{\partial F}{\partial y'} \eta'(x) \right] dx \tag{3}$$

After an integration by parts of the second term, we arrive at

$$\frac{dI}{d\epsilon} = \int_{x_1}^{x_2} \frac{\partial F}{\partial y} \eta(x) dx + \frac{\partial F}{\partial y'} \eta(x) \bigg|_{x_1}^{x_2} - \int_{x_1}^{x_2} \frac{d}{dx}\left(\frac{\partial F}{\partial y'}\right) \eta(x) dx \tag{4}$$

We now use the boundary conditions satisfied by the variational, i.e. $\eta(x_1) = \eta(x_2) = 0$ to remove the boundary term, and arrive at

$$\frac{dI}{d\epsilon} = \int_{x_1}^{x_2} \left[\frac{\partial F}{\partial y} - \frac{d}{dx}\left(\frac{\partial F}{\partial y'}\right) \right] \eta(x) dx$$
$$= 0 \tag{5}$$

This integral vanishes for **arbitrary** variations $\eta(x)$ when the expression in the square brackets vanishes. The resulting equation is the **Euler's condition**

$$\boxed{\frac{\partial F}{\partial y} - \frac{d}{dx}\left(\frac{\partial F}{\partial y'}\right) = 0} \tag{6}$$

where $F = F(x, y(x), y'(x))$. This equation is solved by the stationary curve $y_0(x)$.

Variational Notation: An alternative approach is using the variational symbol δ, which denotes differentiation with respect to an arbitrary variation. The variations of the dependent variables are defined as

$$\delta y(x) \equiv \epsilon \eta, \qquad \delta y'(x) = \frac{d}{dx}\delta y \equiv \epsilon \eta'.$$

The variation of the functional I is defined from the derivative with respect to ϵ as

$$\delta I \equiv \frac{dI}{d\epsilon}d\epsilon,$$

such that the stationarity condition is $\delta I|_{y=y_0} = 0$. Similarly, by the chain rule of differentiation, the variation of the integrand function $F(x, y, y')$ is

$$\delta F \equiv \frac{\partial F}{\partial y}\delta y + \frac{\partial F}{\partial y'}\delta y'.$$

We can repeat the previous derivation using this variational notation. The integral expression of the variational of I follows as

$$\begin{aligned}
\delta I &= \int_{x_1}^{x_2} \delta F\, dx \\
&= \int_{x_1}^{x_2} \left[\frac{\partial F}{\partial y}\delta y + \frac{\partial F}{\partial y'}\delta y'\right] dx \\
&= \int_{x_1}^{x_2} \left[\frac{\partial F}{\partial y} - \frac{d}{dx}\frac{\partial F}{\partial y'}\right] \delta y(x)\, dx \\
&= \int_{x_1}^{x_2} \frac{\delta I}{\delta y}\delta y(x)\, dx
\end{aligned} \qquad (7)$$

where the **functional derivative** of I with respect to $y(x)$ is defined as (for this type of functionals)

$$\frac{\delta I}{\delta y} \equiv \frac{\partial F}{\partial y} - \frac{d}{dx}\frac{\partial F}{\partial y'}.$$

Thus, the stationarity condition corresponds to the functional derivative being equal to zero,

$$\boxed{\frac{\delta I}{\delta y} = 0 \longleftrightarrow \frac{\partial F}{\partial y} - \frac{d}{dx}\left(\frac{\partial F}{\partial y'}\right) = 0.} \qquad (8)$$

The curve parameterization $y(x)$ becomes especially useful when $\partial F/\partial y = 0$, simplifying the equation above to the **first integral** of Euler's condition:

1 Lecture 10: Calculus of Variations: Euler Stationarity Condition

$$\boxed{\frac{d}{dx}\left(\frac{\partial F}{\partial y'}\right) = 0 \rightarrow \frac{\partial F}{\partial y'} = \text{const.}} \quad (9)$$

This form is generally easier to solve.

It is important to note that we have the choice to represent the curve in different ways, such as $y = y(x)$, $x = x(y)$, $\theta = \theta(r)$, or using a parameterization like $(x(t), y(t))$. To find $y_0(x)$, we may choose the curve representation that reduces the Euler's equation into its first integral form.

Example 1 Find the curve with the minimum distance between two points (x_1, y_1) and (x_2, y_2) in the plane.

Solution: The total distance, as a functional, corresponds to the integral over the infinitesimal distance

$$ds = \sqrt{dx^2 + dy^2} = \sqrt{1 + (y')^2} dx$$

such that the total length L as a functional of the curve $y(x)$ and $y'(x)$ can be written as

$$L[y'] = \int_0^L ds = \int_{x_1}^{x_2} \sqrt{1 + (y')^2} dx.$$

We notice that our integrand function $F(y') = \sqrt{1 + (y')^2}$ is independent of both x and y. Applying first integral of the Euler equation, we have that

$$\frac{y'}{\sqrt{1 + (y')^2}} = \kappa,$$

where κ is an integration constant. Equivalently, this means that

$$y' = \frac{\kappa}{\sqrt{1 - \kappa}} = c_1$$

where c_1 is a constant. This is the equation of the straight line

$$y(x) = c_1 x + c_2$$

with the integration constants c_1 and c_2 determined by the boundary conditions. This is a simple problem, but this formalism can be applied elegantly to more complex problems.

1.1.2 Curve Representation $x = x(y)$

The Euler's equation may often reduce to its first integral form when we use $x = x(y)$ as the dependent variable and integrate over y, instead of x. For this, the functional is

$$I[x, x'] = \int_{y_1}^{y_2} F(y, x(y), x'(y)) dy \tag{10}$$

with the corresponding stationarity condition

$$\boxed{\frac{\partial F}{\partial x} - \frac{d}{dy}\left(\frac{\partial F}{\partial x'}\right) = 0} \tag{11}$$

We can go from one formulation to the other through the transformation of variables

$$dx = dy/y' = x' dy.$$

When F is not explicitly dependent on x, i.e. $\partial F/\partial x = 0$, then the parametrization $x = x(y)$ reduces Eq. 11 to its *first integral* form

$$\boxed{\frac{d}{dy}\left(\frac{\partial F}{\partial x'}\right) = 0 \rightarrow \frac{\partial F}{\partial x'} = \text{const.}} \tag{12}$$

Example 2 (*Soap film between two wire rings*) This is a classic problem in finding surfaces of revolution. It is a simplified version of Plateau's problem, which involves determining the surface with the minimum area bounded by given curves.

Solution: The free energy of the soap film is proportional to twice the product of the surface tension and the area of the film (due to the two liquid-air interfaces). Consequently, the equilibrium surface is the one with the minimum area. Given the axial symmetry between the two concentric rings, the surface with the minimum area will be a surface of revolution about the x-axis. Thus, when rotating about the x-axis, the area of the surface of revolution is given by

$$S = 2\pi \int y \, ds = 2\pi \int_{x_1}^{x_2} y\sqrt{1 + (y')^2} \, dx$$

and this is the functional that we want to minimize to find the equilibrium profile $y(x)$. We notice that the integrand

$$F(y, y') = 2\pi y \sqrt{1 + (y')^2}$$

is not explicitly dependent on x, i.e. $\partial F/\partial x = 0$. Hence, we choose to parameterize the profile as $x(y)$ and rewrite the area as an integral over y,

1 Lecture 10: Calculus of Variations: Euler Stationarity Condition

$$S = 2\pi \int_{y_1}^{y_2} y\sqrt{1 + (x')^2}\, dy.$$

The corresponding first integral of the Euler's equation is given by

$$\frac{yx'}{\sqrt{1 + (x')^2}} = \kappa_1,$$

where κ_1 is an arbitrary constant fixed by the boundary conditions. This is a first order differential equation in $x(y)$ as rewritten equivalently,

$$\frac{dx}{dy} = \frac{\kappa_1}{\sqrt{y^2 - \kappa_1^2}}.$$

We can solve it by the separation of variable method, which leads to

$$\int dx = \kappa_1 \int dy\, \frac{1}{\sqrt{y^2/\kappa_1^2 - 1}}.$$

We now use the substitution $y = \kappa_1 \cosh(t)$ such that $dy = \kappa_1 \sinh(t)dt$. Then, the integral reduces to

$$\int dx = \kappa_1 \int dt \to t = \frac{x}{\kappa_1} + \kappa_2,$$

Using the $y(t)$ substitution, we find the generic equilibrium profile

$$y(x) = \kappa_1 \cosh\left(\frac{x}{\kappa_1} + \kappa_2\right),$$

with κ_1 and κ_2 determined by boundary conditions. This is called of the *catenary* equation. For the particular values of $\kappa_1 = 1$ and $\kappa_2 = 0$, we obtained the catenary curve

$$y(x) = \cosh(x).$$

This is shown in Fig. 1. The corresponding surface of revolution is called the catenoid surface and is illustrated in Fig. 2, for two circles centered at $x_{1,2} = \pm 1$ and of radius $y_1 = y_2 = \cosh(1)$.

Quick Primer on the Area of a Surface of Revolution: A surface of revolution is generated by rotating a profile curve $y = y(x)$ around the x-axis. A surface element dS represents a small band rotating around the x-axis, with its thickness determined by the arc length at a specific point on the profile. This surface element can be approximated as an infinitesimal cylinder, where the height is the arclength of the profile at a given point, $ds = \sqrt{1 + [y'(x)]^2}\, dx$, and the circular base has a radius of $y(x)$. Therefore, the area element corresponds to the area of this infinitesimal

Fig. 1 The equilibrium profile $y(x)$ of the soap film

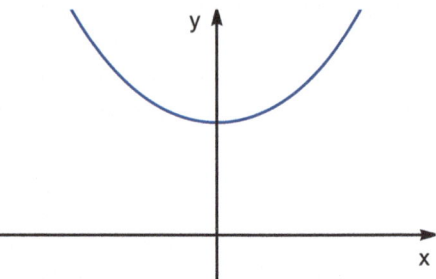

Fig. 2 Surface of revolution obtain by rotating the profile $y(x)$ about the x-axis

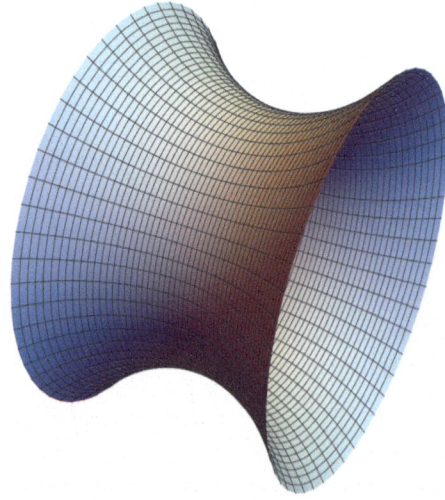

cylinder,
$$dS = 2\pi y(x)ds = 2\pi y(x)\sqrt{1+[y'(x)]^2}dx,$$

such that the total area is the integral over all of these area elements along the surface profile, hence
$$S = \int dS = 2\pi \int_{x_1}^{x_2} y(x)\sqrt{1+[y'(x)]^2}dx.$$

2 Lecture 11: Cycloids and Geodesics

In this lecture, we will apply the stationarity condition to the classical brachistochrone problem and to geodesics. We will make use of the appropriate curve parameterization to reduce the Euler's equation to its first integral form.

2.1 Curve Representation $x = x(y)$

Brachistochrone problem: We aim to find the path of *shortest time* for a particle under gravity and moving between two points (x_1, y_1) and (x_2, y_2). The term "Brachistochrone" is derived from the Greek words brákhistos, meaning shortest, and khrónos, meaning time. Remarkably, the trajectory traversed in the shortest time is the path traced by a point on the circumference of a rolling circle, and is known as the cycloid (shown to Fig. 3). This is, for example, the path traced by a speck of dust attached to a spinning bicycle wheel.

The functional that we want to minimise is the total time spent between the two end points, namely

$$T = \int_0^T dt.$$

We want to use an appropriate integration variable such that $y(x)$ is the dependent variable for $T[y, y']$. For this, we use that the curve results from an equation of motion of a particle falling under gravity. The infinitesimal arclength ds at a given point on the curve relates to the speed at that point as,

$$ds = v(x, y(x))dt$$

Geometrically, the arclength can also be expressed as

$$ds = \sqrt{dx^2 + dy^2} = \sqrt{1 + (y')^2}dx$$

By combing these two expressions, we find the transformation of variables

$$dt = \frac{\sqrt{1 + y'^2}}{v(x, y(x))}dx.$$

In the absence dissipative forces (such as friction), the total energy is conserved. Let us place our coordinate system such that (x_1, y_1) is the origin and the ground zero relative to which we measure the gravitational potential. Thus, the initial total energy of a particle is zero (no kinetic and potential energy) and it must stay zero at any

Fig. 3 Cycloid curve

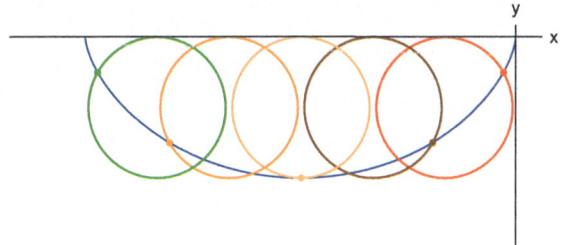

later time. Therefore, at a given time, the potential and kinetic energy must balance each other out

$$\frac{1}{2}mv^2 - mgy = 0 \to v = \sqrt{2gy}.$$

Now, we are ready to construct our functional for the total time spent on the curve as,

$$T[y, y'] = \int dt = \frac{1}{\sqrt{2g}} \int_{x_1}^{x_2} \sqrt{\frac{1+[y'(x)]^2}{y(x)}} dx.$$

The integrand function

$$F(y(x), y'(x)) = \sqrt{\frac{1+[y'(x)]^2}{y(x)}}$$

depends on both $y(x)$ and $y'(x)$ and does not depend explicitly on x. It is harder to solve Euler's equation for this curve parameterization. However, if we let x be the dependent variable $x = x(y)$ and integrate over y, then we can use the first integral for the Euler's equation. For this, we use the change of variables

$$dy = x'dx, \qquad y' = \frac{dy}{dx} = \frac{1}{x'},$$

such that the function can be expressed equivalently as

$$T[y, x'] = \frac{1}{\sqrt{2g}} \int_{y_1}^{y_2} \sqrt{\frac{1+[x'(y)]^2}{y}} dy$$

which is independent of x. The corresponding Euler equation reduces to its first integral form

$$\frac{d}{dy}\left(\frac{\partial F}{\partial x'}\right) = 0 \to \frac{x'}{\sqrt{y(1+x'^2)}} = \sqrt{c},$$

where c is a constant of integration. Solving for x', we find that

$$x' = \frac{dx}{dy} = \sqrt{\frac{cy}{1-cy}}.$$

The generic solution of this first order differential equation can be determined by the separation of variables

$$dx = \sqrt{\frac{cy}{1-cy}} dy.$$

2 Lecture 11: Cycloids and Geodesics

We use the substitution $cy = \sin^2\left(\frac{\theta}{2}\right) = \frac{1}{2}(1 - \cos\theta)$ with $cdy = \sin\left(\frac{\theta}{2}\right)\cos\left(\frac{\theta}{2}\right)d\theta$ such that

$$dx = \frac{\sin\left(\frac{\theta}{2}\right)}{\cos\left(\frac{\theta}{2}\right)}\frac{1}{c}\sin\left(\frac{\theta}{2}\right)\cos\left(\frac{\theta}{2}\right)d\theta$$
$$= \frac{1}{c}\sin^2\left(\frac{\theta}{2}\right)d\theta,$$

from which it follows that

$$dx = \int dx = \frac{1}{2c}\int (1 - \cos\theta)d\theta$$
$$x = \frac{1}{2c}(\theta - \sin\theta) + d,$$

where d is the other integration constant. This is set to 0 by the initial condition that $x(\theta = 0) = x_1 = 0$. Thus, for an arbitrary c, we obtain a family of parameterised curves in terms of θ,

$$x(\theta) = \frac{1}{2c}(\theta - \sin\theta) \tag{13}$$
$$y(\theta) = \frac{1}{c}\sin^2\left(\frac{\theta}{2}\right) = \frac{1}{2c}(1 - \cos\theta), \tag{14}$$

which is also called the family of cycloids. The parameter $a = 1/(2c)$ fixes the radius of the rolling circle. Looking at the parametric equations above for a point on the cycloid, we may also see that this a point on cycle rolling

$$[x(\theta) - a\theta]^2 + [y(\theta) - a]^2 = a^2,$$

as illustrated also in Fig. 4. The brachistochrone curve is a specific kind of cycloid where the circle is rolling under the x-axis as shown in Fig. 3.

Fig. 4 The x and y coordinates of a point on the circle rolling to towards the right

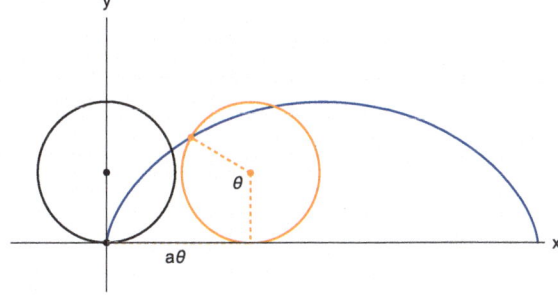

2.2 Polar Representation $\theta = \theta(r)$, $r = r(\theta)$

Problems that have rotational symmetry about an axis are easier to handle when we use polar coordinates. For such problems, the curve is represented as $\theta = \theta(r)$, such that the functional has the polar angle as its main dependent variable,

$$I[\theta, \theta'] = \int_{r_1}^{r_2} F(r, \theta(r), \theta'(r)) dr \tag{15}$$

where $\theta' = d\theta/dr$. The corresponding Euler's equation for this stationary curve is

$$\boxed{\frac{\partial F}{\partial \theta} - \frac{d}{dr}\left(\frac{\partial F}{\partial \theta'}\right) = 0 \quad \theta = \theta(r)} \tag{16}$$

When the integrand is independent of θ, i.e. $\partial F/\partial \theta = 0$, Eq. 16 reduces to its *first integral* form given by

$$\boxed{\frac{\partial F}{\partial \theta'} = \text{const.}} \tag{17}$$

Similarly, for the curve representation $r = r(\theta)$, the Euler's equation for the corresponding $I[r, r']$ follows as

$$\boxed{\frac{\partial F}{\partial r} - \frac{d}{d\theta}\left(\frac{\partial F}{\partial r'}\right) = 0, \quad r = r(\theta)} \tag{18}$$

with its first integral form

$$\boxed{\frac{\partial F}{\partial r'} = \text{const.}} \tag{19}$$

This polar representation is often useful when finding geodesics. We illustrate how to derive parametric equations for geodesics on a cylinder and on a cone using polar representation and the corresponding first integral form.

2.3 Geodesics on Quadratic Surfaces

A quadratic surface is described by a quadratic equation of the surface coordinates (x, y, z). Examples of quadratic surfaces include:

1. Ellipsoid has the normal quadratic form

$$\frac{x^2}{a^2} + \frac{y^2}{b^2} + \frac{z^2}{c^2} = 1.$$

2 Lecture 11: Cycloids and Geodesics

The special case of $a = b = c = R$ corresponds to the sphere.
2. Cone has the normal quadratic form

$$\frac{x^2}{a^2} + \frac{y^2}{b^2} - \frac{z^2}{c^2} = 0.$$

3. Cylinder with the normal quadratic form

$$\frac{x^2}{a^2} + \frac{y^2}{b^2} = 1.$$

4. Hyperboloid of one sheet with the normal quadratic form

$$\frac{x^2}{a^2} + \frac{y^2}{b^2} - \frac{z^2}{c^2} = 1.$$

By minimising the distance between two points on a quadratic surface, we can find its corresponding geodesic. We will consider few examples.

Geodesic on a Cylinder: Let us find the geodesics on the surface of a circular cylinder corresponding to

$$x = R\cos(\theta), \qquad y = R\sin\theta, \qquad z = z$$

where R is the fixed radius of the circular base.

The infinitesimal arclength of a curve on the surface of the cylinder is determined as

$$dl^2 = R^2 d\theta^2 + dz^2$$

thus,

$$dl = \sqrt{R^2 + z'^2}\, d\theta$$

where $z' = dz/d\theta$. The total distance between two points on the cylinder is then

$$L = \int dl = \int d\theta \sqrt{R^2 + (z')^2}.$$

We notice that the integrand $F(z') = \sqrt{R + (z')^2}$ is only a function of z', and hence the corresponding Euler's equation (Eq. 18) simplifies to

$$\frac{d}{d\theta}\frac{\partial F}{\partial z'} = 0,$$

from implies that

$$\frac{z'}{\sqrt{R^2 + z'^2}} = constant.$$

Since R is also a constant, this is equivalent to

$$z' = a$$

which has the helix as its solution

$$z(\theta) = a\theta + b \tag{20}$$

where a and b are integration constants. Figure 5 show a helix on the cylinder for $a = 1$ and $b = 0$.

Obs: Notice that $a = 0$ corresponds to a curve with constant $z = b$, which is an circular arc in the $x - y$ plane along the circumference of the cylinder. By rewriting the geodesic equation as $\theta(z)$ instead

$$\theta(z) = \tilde{a}z + \tilde{b},$$

and setting $\tilde{a} = 0$, we obtain the curve of constant azimuthal angle $\theta = \tilde{b}$ corresponding to a straight line segment on the surface of the cylinder and parallel to the z-axis.

Geodesic on a Cone: Let us now find the geodesics on the surface of this cone

$$z = \sqrt{x^2 + y^2}.$$

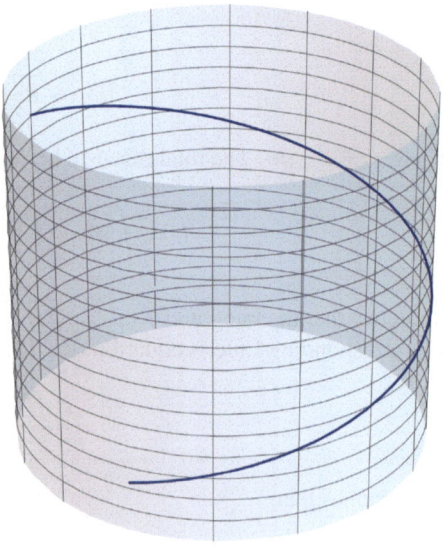

Fig. 5 Example of a helix on the cylinder for $a = 1$ and $b = 0$

2 Lecture 11: Cycloids and Geodesics

We can use the cone symmetry around the z axis and choose the polar coordinates in the (x, y)-plane with $r = \sqrt{x^2 + y^2}$ being the radius of the circle. Thus, an infinitesimal arclength of a curve on the surface of the cone is determined as

$$dl^2 = ds^2 + dz^2 = dr^2 + r^2 d\theta^2 + dz^2$$

Using the normal form $z = r$, we obtain

$$dl = \sqrt{2dr^2 + r^2 d\theta^2} = \sqrt{2 + r^2(\theta')^2} dr.$$

To find the cone geodesics, we want to minimize the total length of a curve between two points given by

$$L = \int dl = \int_{r_1}^{r_2} \sqrt{2 + r^2(\theta')^2} dr.$$

Since the functional depends only on θ', we can use the first integral form given by

$$\frac{\partial}{\partial \theta'} \sqrt{2 + r^2(\theta')^2} = \frac{r^2 \theta'}{\sqrt{2 + r^2(\theta')^2}} = \kappa$$

where κ is the integration constant. Hence, $\theta(r)$ satisfies this differential equation

$$\frac{d\theta}{dr} = \frac{1}{r} \frac{\sqrt{2}\kappa}{\sqrt{r^2 - \kappa^2}},$$

which we can integrate as

$$\int d\theta = \sqrt{2} \int dr \frac{\kappa}{r^2} \frac{1}{\sqrt{1 - \kappa^2/r^2}}.$$

Using the change of variables $\omega = \kappa/r < 1$, we write the integral on the left hand side as

$$\int d\theta = -\sqrt{2} \int d\omega \frac{1}{\sqrt{1 - \omega^2}}.$$

Furthermore, we substitute $\omega = \cos \alpha$ such that

$$\int d\theta = \sqrt{2} \int d\alpha \frac{\sin \alpha}{\sqrt{1 - \cos^2 \alpha}} = \sqrt{2} \int d\alpha$$

and

$$\theta + \theta_0 = \sqrt{2} \arccos \left(\frac{\kappa}{r} \right)$$

or equivalently,

$$r\cos\left(\frac{\theta+\theta_0}{\sqrt{2}}\right)=\kappa, \quad r=\kappa\sec\left(\frac{\theta+\theta_0}{\sqrt{2}}\right),$$

where θ_0 is an integration constant. Both κ and θ_0 are fixed by the boundary conditions. The parametric equations for the cone geodesic are given by

$$x = \kappa \sec\left(\frac{\theta+\theta_0}{\sqrt{2}}\right)\cos\theta$$
$$y = \kappa \sec\left(\frac{\theta+\theta_0}{\sqrt{2}}\right)\sin\theta$$
$$z = \kappa \sec\left(\frac{\theta+\theta_0}{\sqrt{2}}\right). \tag{21}$$

Figure 6 shows a geodesic corresponding to $\theta_0 = 0$ and $\kappa = 1$.

3 Lecture 12: Fermat's Principle and Hamilton's Principle

In optics, variational calculus is fundamental to Fermat's principle, which states that light rays follow paths that minimize travel time through an optical medium. Similarly, in classical mechanics, Hamilton's principle of stationary action underpins Newton's laws of motion.

In this lecture, we will apply the Euler's condition to determine the minimum-time path of a light ray. We derive the classical Snell's law of refraction using the variational approach. We will also use the Euler condition associated with stationary action to derive the Euler-Lagrange equations of motion for Hamiltonian systems.

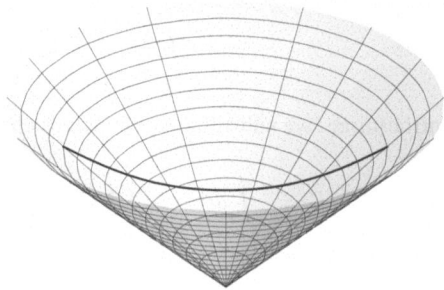

Fig. 6 Example of a cone geodesic for $\theta_0 = 0$ and $\kappa = 1$

3 Lecture 12: Fermat's Principle and Hamilton's Principle

3.1 Fermat's Principle

Fermat's principle postulates that a ray of light travels in an optical medium on the path of least time. An optical medium is characterized by its **index of refraction** or **refractive index**, $n(\mathbf{r})$. This is inversely proportional to the speed of light in an optical medium

$$v = c/n,$$

where c is the speed of light in the vacuum i.e. $n_{vacuum} = 1$. Thus, in any other optical medium, light travels at a lower speed. It is fascinating that nowadays we can slow down light and almost trap it inside a cloud of ultra cold atoms, such as Bose Einstein condensates which has a refraction index of about 10^7 corresponding to a light speed of about $v \sim 30$ m/s thus comparable with the car speed on the highway.

3.1.1 Snell's Law

Let us consider a light ray traveling in the (x, y) plane in a layered optical medium, such that the refractive index $n(x)$ depends on x and is uniform in the y direction.

We will show that the Snell's law of refraction is the Euler's condition that follows from Fermat's principle. The path of the light ray is a curve $y(x)$ which minimises the total time spent on the path. The infinitesimal time increment is related to the infinitesimal arclength on the path traversed by light as

$$dt = \frac{ds}{v} = \frac{1}{c}n(x)\sqrt{1+y'^2}dx.$$

Thus, the total time is a functional with dependent variables y and y' given by

$$T[y, y'] = \int dt = \frac{1}{c}\int n(x)\sqrt{1+y'^2}dx$$

where $y' = dy/dx$. We have chosen the appropriate curve parameterization since the integrand dependents on y' and the integration variable. The corresponding Euler's equation reduces to its first integral form given by

$$\frac{d}{dx}\frac{\partial}{\partial y'}\left[n(x)\sqrt{1+y'^2}\right] = 0,$$

which implies that

$$n(x)\frac{y'}{\sqrt{1+y'^2}} = \kappa,$$

where κ is the integration constant. This relation holds everywhere along the path, i.e. for every x. Now, we consider an interface between two homogeneous media, whereby the refractive index is constant in each medium and jumps abruptly at their interface,

$$n(x) = \begin{cases} n_1, & x > 0 \\ n_2, & x \leq 0 \end{cases} \quad (22)$$

We solve the stationarity condition for each homogeneous side and match the solutions at the interface, i.e. $x = 0$. For $x \neq 0$, the refractive index is a constant, and the solution is a straight line in each half-plane

$$\frac{y'}{\sqrt{1+y'^2}} = a \rightarrow y(x) = ax + b, \quad x \neq 0,$$

where the slope a is determined by the interfacial condition at $x = 0$. The curve is continuous across the interface, while its slope has a jump determined by the jump in the refraction index. Applying the stationarity condition on each side of $x = 0$, we have that

$$n_1 \frac{a_1}{\sqrt{1+a_1^2}} = n_2 \frac{a_2}{\sqrt{1+a_2^2}}.$$

Using that the slopes are tangents: $a_1 = y_1' = \tan(\theta_1)$ and $a_2 = y_2' = \tan(\theta_2)$ with $\theta_{1,2}$ being the angles between the lines meeting at $x = 0$ and x axis (see Fig. 7), the above interfacial condition reduces to the classical **Snell's law**:

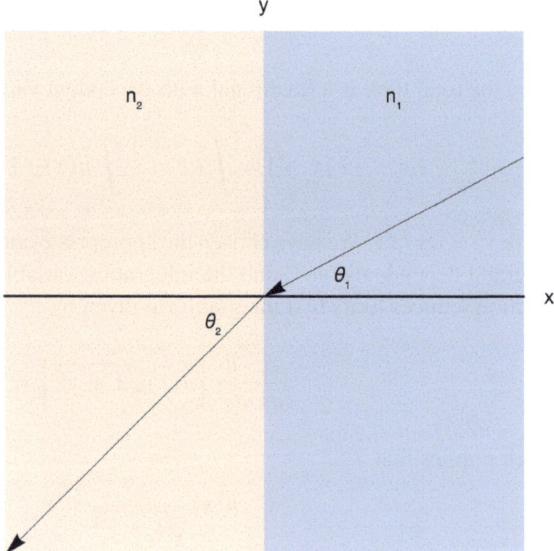

Fig. 7 Snell's law

$$n_1 \frac{\tan \theta_1}{\sqrt{1 + \tan^2 \theta_1}} = n_2 \frac{\tan \theta_1}{\sqrt{1 + \tan^2 \theta_1}} \tag{23}$$

$$n_1 \sin \theta_1 = n_2 \sin \theta_2. \tag{24}$$

When θ_1 is the *incidence angle*, θ_2 is the *refractive angle*.

3.2 Hamilton's Principle

3.2.1 One Coordinate Trajectory $x = x(t)$: Euler-Lagrange Equation

A curve can also represent the trajectory of a particle moving in one-dimension (coordinate trajectory), hence $x = x(t)$. We may introduce the action as the functional

$$S[x, \dot{x}] = \int_{t_1}^{t_2} L(t, x(t), \dot{x}(t)) dt \tag{25}$$

where $\dot{x} = \frac{dx}{dt}$, and $L(t, x, \dot{x})$ is the **Lagrangian** function defined as the **kinetic energy** T *minus* the **potential energy** V, i.e.

$$L = \frac{1}{2} m \dot{x}^2 - V(x).$$

The corresponding stationary condition becomes

$$\boxed{\frac{\partial L}{\partial x} - \frac{d}{dt} \left(\frac{\partial L}{\partial \dot{x}} \right) = 0.} \tag{26}$$

This is known as the Euler-Lagrange equation of motion of one particle moving in the x direction. Using the definition of the Lagrangian, we see that this reduces to the Newton's equation

$$\boxed{m \ddot{x} = -\frac{d}{dx} V(x)} \tag{27}$$

for Hamiltonian dynamics.

When $\partial L / \partial x = 0$, then Eq. 26 reduces the *first integral*

$$\boxed{\frac{\partial L}{\partial \dot{x}} = const. \rightarrow m \dot{x} = const.} \tag{28}$$

which corresponds to the conservation of momentum.

3.2.2 Multiple Trajectories: Euler-Lagrange Equations

This principle of stationary action can be extended to higher spatial dimensions and multiple particles. In the general case of N particles in d dimensions, we often use the generalized position and velocity coordinates, $(q_i, \dot{q}_i)$ where $i = 1, \ldots, n$ labels each component of motion (there are $n = d \times N$ generalised coordinates). The Lagrangian of the system reads as

$$L = \frac{1}{2} \sum_i m \dot{q}_i^2 - V(q_1, \ldots, q_n).$$

For a single point particle moving in 3D, $q_i \equiv x, y, z$ and $\dot{q}_i \equiv \dot{x}, \dot{y}, \dot{z}$.

In general, the **action** of the system is given by the functional

$$S[q_1, q_n, \ldots \dot{q}_1 \ldots \dot{q}_n] = \int_{t_i}^{t_f} L(t, \{q_i, \dot{q}_i\}) dt,$$

which has multiple variables corresponding to each coordinate trajectory.

Let us examine the case of two coordinate trajectories, namely $q_1 = q_1(t)$ and $q_2 = q_2(t)$, which may correspond either to one point particle in t2D or of two point particles in 1D. We consider *independent* variations along each trajectory, i.e. δq_1 and δq_2, such that the end points on each trajectory remain fixed. The stationary condition of the action implies that variational of action must vanish under such infinitesimal variations

$$\delta S[q_1, \dot{q}_1, q_2, \dot{q}_2] = 0.$$

By carrying the calculus of variations for the two independent coordinates, we have that

$$\delta S[q_1, \dot{q}_1, q_2, \dot{q}_2] = \int dt \left[\frac{\partial L}{\partial q_1} \delta q_1 + \frac{\partial L}{\partial \dot{q}_1} \delta \dot{q}_1 + \frac{\partial L}{\partial q_2} \delta q_2 + \frac{\partial L}{\partial \dot{q}_2} \delta \dot{q}_2 \right]$$
$$= \int dt \left[\left(\frac{\partial L}{\partial q_1} - \frac{d}{dt} \frac{\partial L}{\partial \dot{q}_1} \right) \delta q_1 + \left(\frac{\partial L}{\partial q_2} - \frac{d}{dt} \frac{\partial L}{\partial \dot{q}_2} \right) \delta q_2 \right]$$
$$= 0$$

Using that δq_1 and δq_2 are *independent* variations, we find that each integrand must vanish independently, Hence, each coordinate satisfies its own Euler condition:

$$\frac{d}{dt} \frac{\partial L}{\partial \dot{q}_1} - \frac{\partial L}{\partial q_1} = 0 \qquad (29)$$

$$\frac{d}{dt} \frac{\partial L}{\partial \dot{q}_2} - \frac{\partial L}{\partial q_2} = 0 \qquad (30)$$

3 Lecture 12: Fermat's Principle and Hamilton's Principle

The Lagrangian function for this system reads as

$$L(q_1, q_2, \dot{q}_1, \dot{q}_2) = T(\dot{q}_1, \dot{q}_2) - V(q_1, q_2) = \frac{1}{2}m_1\dot{q}_1^2 + \frac{1}{2}m_2\dot{q}_2^2 - V(q_1, q_2).$$

Inserting it into Eqs. 29 and 30, we arrive at the Newton's equation of motion under potential forces

$$m\frac{d^2q_1}{dt^2} = -\frac{\partial V}{\partial q_1}$$
$$m\frac{d^2q_2}{dt^2} = -\frac{\partial V}{\partial q_2}.$$

Example 3 Let us derive the equations of motion for one point particle of mass m and moving in the plane. The potential field depends only on the distance from the origin of the moving particle $V(r)$, that is rotationally symmetric. The kinetic energy follows from v^2 which can be determined from the infinitesimal arclength $ds^2 = dr^2 + r^2 d\theta^2$ along the particle's trajectory, i.e.

$$v^2 = \left(\frac{ds}{dt}\right)^2 = \left(\frac{dr}{dt}\right)^2 + r^2\left(\frac{d\theta}{dt}\right)^2 = \dot{r}^2 + r^2\dot{\theta}^2,$$

where the first component relates to the radial speed and the second one to the angular speed. The Lagrangian function at a given point on the trajectory is

$$L = \frac{1}{2}mv^2 - V(r) = \frac{1}{2}m(\dot{r}^2 + r^2\dot{\theta}^2) - V(r),$$

such that the action functional is

$$S[r, \theta, \dot{r}, \dot{\theta}] = \int_0^T dt\left[\frac{1}{2}m(\dot{r}^2 + r^2\dot{\theta}^2) - V(r)\right].$$

The corresponding Euler-Lagrange Eqs. 29–30 are

$$\frac{d}{dt}(m\dot{r}) - mr\dot{\theta}^2 + V'(r) = 0 \tag{31}$$

$$\frac{d}{dt}(mr^2\dot{\theta}) = 0, \tag{32}$$

where $V'(r) = dV(r)/dr$. Equivalently, this is

$$m(\ddot{r} - r\dot{\theta}^2) = -V'(r) \tag{33}$$

$$\frac{d}{dt}(mr^2\dot{\theta}) = 0. \tag{34}$$

We may recognize the second equation as being the conservation of angular momentum, $l = mr^2\dot{\theta}$, which is due to the rotational symmetry of the potential and the Hamiltonian dynamics (no dissipation).

There is a fundamental connection between symmetry and conservation laws for Hamiltonian systems, as first recognised by Emmy Noether:

$$\boxed{\text{Symmetry} \iff \text{Conservation Law}}$$

In the previous example, the rotational invariance translates into the conservation of angular momentum under Hamiltonian dynamics.

$$\text{Rotational Symmetry} \iff \text{Conservation of Angular Momentum}$$

Similarly, a system that has *translational symmetry* will conserve *momentum*.

$$\text{Translational Symmetry} \iff \text{Conservation of Momentum}$$

A system that is invariant under *time translations* conserves the *total energy* (Hamiltonian).

$$\text{Time Symmetry} \iff \text{Conservation of Energy}$$

In Noether's approach, we can determine the conservation law directly from the variational of the action integral. We do so by taking into account the symmetries of the Lagrangian without having to derive all the equations of motion. We illustrate this technique for the previous example of a point particle moving under a rotationally-symmetric force. Let us look again at the corresponding action integral

$$S[r, \theta, \dot{r}, \dot{\theta}] = \int_0^T dt \left[\frac{1}{2}m(\dot{r}^2 + r^2\dot{\theta}^2) - V(r)\right],$$

Notice that the Lagrangian remains unchanged by a rotation with an *arbitrary constant* phase

$$\theta(t) \to \theta(t) + \epsilon\alpha$$

For a stationary action the requirement is that the Lagrangian remains unchanged by an arbitrary rotational with any time-dependent phase, including this one that is inspired by the rotational symmetry

$$\theta(t) \to \theta(t) + \epsilon(t)\alpha.$$

Thus, we may relate the variational of θ with the rotation rate α

$$\delta\theta(t) \equiv \epsilon(t)\alpha.$$

3 Lecture 12: Fermat's Principle and Hamilton's Principle

Inserting, this into the variational of the action and applying the integration by parts (with vanishing boundary terms), we find that

$$\delta S = \int_0^T dt\, mr^2\dot{\theta}\delta\dot{\theta}$$
$$= \alpha \int_0^T dt\, mr^2\dot{\theta}\dot{\epsilon}$$
$$= -\alpha \int_0^T dt\, \frac{d}{dt}(mr^2\dot{\theta})\epsilon$$
$$= 0.$$

The stationarity condition $\delta S = 0$ for an arbitrary ϵ immediately results in the conservation law of angular momentum

$$\frac{d}{dt}(mr^2\dot{\theta}) = 0.$$

Notice that in this example the Lagrangian is also not explicitly dependent on time, i.e.

$$\frac{\partial L}{\partial t} = 0$$

This means that it is invariant under time translations, $t \to t + \epsilon$, with an *arbitrary constant* ϵ. It implies that the Lagrangian remains unchanged to the following transformation of the variable q:

$$q(t) \to q(t + \epsilon)$$

For an infinitesimal ϵ, we can Taylor expand to first order in ϵ such that the above transformation reduces to a global translation of the coordinate variable

$$q(t) \to q(t) + \epsilon\dot{q}$$

This leads us to consider variations that mimics the invariant transformation by making ϵ time-dependent:

$$\delta q \equiv \epsilon(t)\dot{q}$$

The variational calculus of the action with this variation is

$$\delta S = \int dt \left[\frac{\partial L}{\partial q}\delta q + \frac{\partial L}{\partial \dot{q}}\delta \dot{q} \right] \tag{35}$$

$$= \int dt \left[\frac{\partial L}{\partial q}\epsilon\dot{q} + \frac{\partial L}{\partial \dot{q}}\frac{d}{dt}(\epsilon\dot{q}) \right] \tag{36}$$

$$= \int dt \left[\left(\frac{\partial L}{\partial q} \dot{q} + \frac{\partial L}{\partial \dot{q}} \ddot{q} \right) \epsilon + \frac{\partial L}{\partial \dot{q}} \dot{\epsilon} \dot{q} \right] \tag{37}$$

The first term in the parenthesis represents the total time derivative,

$$\frac{dL}{dt} = \frac{\partial L}{\partial q} \dot{q} + \frac{\partial L}{\partial \dot{q}} \ddot{q},$$

using that $\frac{\partial L}{\partial t} = 0$. Thus, the action can be rewritten, after an integration by parts of the last term, as

$$\delta S = \int dt \left[\frac{dL}{dt} - \frac{d}{dt} \left(\frac{\partial L}{\partial \dot{q}} \dot{q} \right) \right] \epsilon \tag{38}$$
$$= 0 \tag{39}$$

Henceforth, time invariance leads to the conservation of the Hamiltonian

$$\frac{d}{dt} \left[\dot{q} \frac{\partial L}{\partial \dot{q}} - L \right] = 0 \iff \frac{d}{dt} H = 0 \tag{40}$$

where the Hamiltonian is defined as the conjugate of the Lagrangian by the Legendre transform

$$H = \dot{q} \frac{\partial L}{\partial \dot{q}} - L = T(\dot{q}) + V(q).$$

Ordinary Differential Equations

1 Lecture 13: First Order ODEs

In this lecture, we will introduce ordinary differential equations and discuss different methods for solving 1st and 2nd order equations.

Definition 1 An ordinary differential equation (ode) is a general expression containing derivatives of a function of one variable, $y = y(x)$

$$F(x, y, y', y'', y''', \ldots y^{(n)}) = 0,$$

where $y' = dy/dx$, $y'' = d^2y/dx^2, \ldots y^{(n)} = d^n y/dx^n$ are derivatives of $y(x)$.

Definition 2 The order of an ode is given by the highest derivative of $y(x)$.

Example 1 $y'' + x^2 y' + y^2 = 5$ is an ODE of order 2

Definition 3 Linear ode's have a general form

$$a_0 y + a_1 y' + a_2 y'' + a_3 y''' + \cdots + a_n y^{(n)} = b,$$

where $a_0, a_1 \ldots, a_n$ and b are **constants** or **functions of** x. Each term is a linear function of $y(x)$ and its derivatives.

Linear ode's have **general** solutions or **family** of solutions, which have n *arbitrary* constants, n is the **order** of the ode.

Boundary or Initial Conditions are required to find the *specific* solution fulfilling these conditions. For a unique specific solution, the number of boundary conditions must equal the number of coefficients determined by the order of the ODE. Conditions imposed at $x = 0$ are also called **initial conditions**.

The original version of the chapter has been revised. A correction to this chapter can be found at https://doi.org/10.1007/978-3-031-77053-1_7

© The Author(s), under exclusive license to Springer Nature Switzerland AG 2025, corrected publication 2025 L. Angheluta, *Analytical Methods in Physics*, https://doi.org/10.1007/978-3-031-77053-1_3

There are several generic methods that work for certain kinds of ode's of a given order. In this lecture, we will explore these methods and apply them to first and second-order ode's.

1.1 First Order ODE's

First order differential equations are expressions that link the variable x and the function $y(x)$ with the first derivative of $y(x)$

$$F(x, y, y') = 0.$$

The ode is a **linear** superposition of y and y' in the general form

$$y' + P(x)y = Q(x).$$

We will cover the basic techniques for solving first-order ode's, including the integrating factor method, which is specific to linear equations, as well as two general methods that can be applied to both linear and certain types of nonlinear first-order ode's.

1.1.1 Linear 1st Order Ode

The generic form of a **linear** first order ode is given by

$$y' + P(x)y = Q(x), \qquad (1)$$

where $P(x)$ and $Q(x)$ are known continuous functions of x or constants. The task is to find a general solution of $y(x)$. Let us introduce the anti-derivative of $P(x)$,

$$I(x) = \int_{-\infty}^{x} P(s)ds$$

and multiply both sides of Eq. 1 with the exponential factor $e^{I(x)}$. This results in

$$y'e^{I(x)} + P(x)ye^{I(x)} = Q(x)e^{I(x)}$$
$$\frac{d}{dx}\left[y(x)e^{I(x)}\right] = Q(x)e^{I(x)}$$
$$y(x)e^{I(x)} = \int_{-\infty}^{x} dz Q(z)e^{I(z)} + C,$$

where C is the integration constant. Hence, the general solution can be written as

1 Lecture 13: First Order ODEs

$$y(x) = e^{-I(x)} \int_{-\infty}^{x} dz\, Q(z) e^{I(z)} + C e^{-I(x)}, \qquad I(x) = \int_{-\infty}^{x} P(s)\,ds. \qquad (2)$$

Example 2 Find the general solution of this linear first order ode

$$y' + y = e^x \qquad (3)$$

Solution: Since the coefficient in front of y is 1, we multiply on both sides with e^x,

$$y' e^x + y e^x = e^{2x}$$

$$\frac{d}{dx}\left[y(x) e^x\right] = e^{2x}$$

$$y(x) e^x = \frac{1}{2} e^{2x} + C,$$

where C is the integration constant. Hence, the general solution is

$$y(x) = \frac{1}{2} e^x + C e^{-x}. \qquad (4)$$

1.1.2 Method of Separation of Variables

Suppose we now take a first order ode of this general form

$$g(y) y' = f(x) \qquad (5)$$

such that all the y-dependent terms can be separated from the x-dependent terms as

$$g(y)\,dy = f(x)\,dx.$$

Then, the solution $y(x)$ is obtained by integrating each side of the differential form

$$\int g(y)\,dy = \int f(x)\,dx.$$

Example 3 Find the solution of

$$x y' = y$$

Solution: We write it in the differential form and separate the variables as

$$\frac{dy}{y} = \frac{dx}{x},$$

which we can solve when $y \neq 0$ and $x \neq 0$. Integrating on both sides, we get

$$\ln y = \ln x + C \rightarrow y(x) = Ax.$$

where $A = e^C$ is the integration constant. The general solution is a family of straight lines of various slopes that intersect at $y = x = 0$.

1.1.3 Method of Integrating Factors

Let us now consider the generic first order ode

$$Q(x, y)y' + P(x, y) = 0 \tag{6}$$

or, equivalently, in the differential form

$$P(x, y)dx + Q(x, y)dy = 0. \tag{7}$$

When the coefficients $P(x, y)$ and $Q(x, y)$ are functions that satisfy the integrability condition

$$\frac{\partial P}{\partial y} = \frac{\partial Q}{\partial x},$$

then there exists an *exact* differential form of a function $F(x, y)$, i.e.

$$dF = P(x, y)dx + Q(x, y)dy, \tag{8}$$

where

$$Q = \frac{\partial F}{\partial y}, \quad P = \frac{\partial F}{\partial x}.$$

The integrability condition corresponds to the commutation of the derivatives

$$\frac{\partial}{\partial x}\left(\frac{\partial F}{\partial y}\right) = \frac{\partial}{\partial y}\left(\frac{\partial F}{\partial x}\right).$$

The function F is determined by its *integrating factors* P and Q. Namely, the general solution of Eq. (8) is given by solving

$$dF = 0 \rightarrow F(x, y) = const. \rightarrow y(x) \tag{9}$$

The general expression of $F(x, y)$ is obtained by integrating the equations for P and Q.

$$P(x, y) = \frac{\partial F}{\partial x} \rightarrow F(x, y) = \int P(x, y)dx + f(y)$$

$$Q(x, y) = \frac{\partial F}{\partial y} \rightarrow F(x, y) = \int Q(x, y) dy + g(x)$$

Matching the integration factors, we have that

$$F(x, y) = \int P(x, y) dx + \int Q(x, y) dy.$$

Then, the general solution is obtained from

$$\int P(x, y) dx + \int Q(x, y) dy = C.$$

The Separation of Variables is a particular case of the method of integrating factors. That is when the ode has this form

$$Q(y) y' + P(x) = 0 \rightarrow Q(y) dy + P(x) dx = 0 \qquad (10)$$

We also notice that functions $P(x)$ and $Q(y)$ satisfy the integrability condition for an exact differential

$$\frac{\partial P}{\partial y} = \frac{\partial Q}{\partial x} = 0.$$

Example 4 Find the expression satisfied by the general solution $y(x)$ of this nonlinear first order ode

$$(x^3 - y^3) y' + 3x^2 y = 0.$$

Solution: The ode has the differential form

$$dF = 3x^2 y \, dx + (x^3 - y^3) dy = 0,$$

where $F(x, y)$ is the exact differential that satisfies the integrability condition

$$\frac{\partial}{\partial y}(3x^2 y) = \frac{\partial}{\partial x}(x^3 - y^3) = 3x^2.$$

The function $F(x, y)$ is determined by the integrating factors

$$\frac{\partial F}{\partial x} = 3x^2 y \Rightarrow F(x, y) = 3y \int x^2 dx + f(y) = yx^3 + f(y)$$

$$\frac{\partial F}{\partial y} = x^3 - y^3 \Rightarrow F(x, y) = \int (x^3 - y^3) dy + g(x) = yx^3 - \frac{y^4}{4} + g(x).$$

By identifying these two expressions, we find that $F(x, y) = yx^3 - \frac{y^4}{4}$. The general solution $y(x)$ satisfies the equation of constant $F(x, y) = C$, namely

$$yx^3 - \frac{y^4}{4} = C.$$

1.2 Linear Second Order Ode's

The general form of a *linear* second order ode is

$$\boxed{y'' + P(x)y' + Q(x)y = R(x),} \tag{11}$$

where $P(x)$, $Q(x)$ and $R(x)$ are continuous, known functions of x or constants.

Homogeneous ODE i.e. $R(x) = 0$: The **general solution** is given as a linear superposition of two independent solutions $y_1(x)$ and $y_2(x)$

$$\boxed{y_h(x) = c_1 y_1(x) + c_2 y_2(x)} \tag{12}$$

Non-Homogeneous ODE i.e. $R(x) \neq 0$: The **general solution** is given by the general solution of the homogeneous equation from Eq. 12 and a *particular* solution $y_p(x)$ determined by the source term $R(x)$.

$$\boxed{y(x) = y_h(x) + y_p(x).} \tag{13}$$

A *specific* solution is fixed by specializing the arbitrary constants c_1 and c_2 such that the general solution satisfies given **boundary conditions**.

1.2.1 Second Order Ode's with Constant Coefficients

Method of Undetermined Coefficients

Homogeneous ODE $R(x) = 0$: Let us consider a differential equation of this form

$$\boxed{y'' + ay' + by = 0,} \tag{14}$$

The task is to find the two independent solutions for this homogeneous equation. We introduce the *auxiliary* or *characteristic* equation

$$\lambda^2 + a\lambda + b = 0,$$

with the roots

$$\lambda_{1,2} = \frac{-a \pm \sqrt{a^2 - 4b}}{2},$$

1 Lecture 13: First Order ODEs

such that we can factorize the differential equation into product of 1st order differentials as

$$\left(\frac{d}{dx} - \lambda_1\right)\left(\frac{d}{dx} - \lambda_2\right) y = 0.$$

We can solve this by reducing the equation above to a set of two coupled first order ode's,

$$\left(\frac{d}{dx} - \lambda_1\right) u(x) = 0$$

$$\left(\frac{d}{dx} - \lambda_2\right) y(x) = u(x).$$

The first equation has the general solution

$$u(x) = c_1 e^{\lambda_1 x}$$

which implies that the integral solution of the second equation is

$$y(x) = c_2 e^{\lambda_2 x} + c_1 e^{\lambda_2 x} \int ds\, e^{(\lambda_1 - \lambda_2)s}.$$

We have the following two situations:

(a) $\lambda_1 \neq \lambda_2$: We integrate the exponential integrand and obtain, the general solution of $y(x)$ given by

$$\boxed{y(x) = c_1 e^{\lambda_1 x} + c_2 e^{\lambda_2 x}}$$

(b) $\lambda_1 = \lambda_2 = \lambda$: In this case, $e^{(\lambda_1 - \lambda_2)s} = 1$ and its integral is just x. Thus, the general solution reduces to

$$\boxed{y(x) = (c_1 x + c_2) e^{\lambda x}}$$

Example 5 Find the general solution of this equation

$$y'' - 4y' + 3y = 0 \tag{15}$$

Solution: The characteristic equation

$$\lambda^2 - 4\lambda + 3 = 0,$$

has the distinct real roots $\lambda_1 = 1$ and $\lambda_2 = 3$. Thus, the general solution is

$$y(x) = c_1 e^x + c_2 e^{3x}$$

with integration constants determined by boundary conditions.

Example 6 Find the general solution of this equation

$$y'' - 2y' + y = 0. \qquad (16)$$

Solution: The characteristic equation

$$\lambda^2 - 2\lambda + 1 = 0,$$

has equal roots $\lambda = \lambda_{1,2}$. The ode can be written equivalently as:

$$\left(\frac{d}{dx} - 1\right)\left(\frac{d}{dx} - 1\right) y = 0$$

Let us use the substitution

$$\left(\frac{d}{dx} - 1\right) y \equiv u(x),$$

such that

$$\left(\frac{d}{dx} - 1\right) u = 0.$$

The corresponding general solution is

$$u(x) = c_1 e^x.$$

We solve the remaining equation from substitution

$$\left(\frac{d}{dx} - 1\right) y = c_1 e^x$$

which leads to

$$e^{-x} y' - y e^{-x} = c_1 \Rightarrow$$
$$\frac{d}{dx}\left(e^{-x} y\right) = c_1 \rightarrow$$
$$y e^{-x} = c_1 x + c_2.$$

Hence, the general solution follows as

$$y(x) = (c_1 x + c_2) e^x.$$

Example 7 Find the general solution of this equation

$$y'' - 2y' + 2y = 0. \tag{17}$$

Solution: The characteristic equation

$$\lambda^2 - 2\lambda + 2 = 0,$$

has the complex conjugate roots $\lambda_{1,2} = 1 \pm i$. The ode can be written equivalently as:

$$\left(\frac{d}{dx} - 1 + i\right)\left(\frac{d}{dx} - 1 - i\right) y = 0$$

We are solving
$$y' - (1-i)y = 0$$

as

$$\frac{dy}{y} = (1-i)dx \rightarrow$$
$$\ln y = (1-i)x + c \rightarrow$$
$$y(x) = c e^x e^{-ix}.$$

Similarly, the solution of
$$y' - (1+i)y = 0$$

is
$$y(x) = c e^x e^{ix}.$$

Hence, the general solution is a superposition of these two independent solutions,

$$y(x) = e^x (c_1 e^{-ix} + c_2 e^{ix}).$$

2 Lecture 14: Linear Second Order ODEs

2.1 Second Order Ode's with Constant Coefficients

This is a class of ode's that are analytically tractable. We present how to apply the method of undetermined coefficients to find the general solution of non-homogeneous odes.

2.1.1 Method of Undetermined Coefficients

Non-Homogeneous ODE $R(x) \neq 0$: A non-homogeneous, linear second order ode with constant coefficients has the general form

$$\boxed{y'' + ay' + by = R(x)}$$

Recall that the homogeneous solution

$$y_h(x) = c_1 y_1(x) + c_2 y_2(x),$$

is obtained through the characteristic equation $\lambda^2 + a\lambda + b = 0$. The roots $\lambda_{1,2}$ determine the independent solutions y_1 and y_2. In addition, we also have a particular solution $y_p(x)$ that is follows from the *forcing* term $R(x)$. Below, we discuss three classes of forcing terms for which we determine y_p.

The second order ode can be written as the set of first order ode's

$$\left(\frac{d}{dx} - \lambda_1\right) u(x) = R(x)$$

$$\left(\frac{d}{dx} - \lambda_2\right) y(x) = u(x).$$

The corresponding integral solutions follow as

$$u(x) = c_1 e^{\lambda_1 x} + e^{\lambda_1 x} \int^x ds\, R(s) e^{-\lambda_1 s}$$

$$y(x) = c_2 e^{\lambda_2 x} + e^{\lambda_2 x} \int^x ds\, u(s) e^{-\lambda_2 s}$$

$$= c_2 e^{\lambda_2 x} + c_1 e^{\lambda_2 x} \int^x ds\, e^{(\lambda_1 - \lambda_2)s} + e^{\lambda_2 x} \int^x ds\, e^{(\lambda_1 - \lambda_2)s} \int^s dt\, R(t) e^{-\lambda_1 t}$$

$$= y_h(x) + e^{\lambda_2 x} \int^x ds\, e^{(\lambda_1 - \lambda_2)s} \int^s dt\, R(t) e^{-\lambda_1 t}.$$

The integral expression of the particular solution is

$$y_p(x) = e^{\lambda_2 x} \int^x ds\, e^{(\lambda_1 - \lambda_2)s} \int^s dt\, R(t) e^{-\lambda_1 t}.$$

Thus, the form of the particular solution depends on values of $\lambda_{1,2}$ and the sourcing term $R(x)$. Let us take few examples to illustrate how we compute the particular solution in practice.

Example 8 Find the general solution for this ode

$$y'' - 4y' + 3y = 2e^{-x}.$$

Solution: The characteristic equation

$$\lambda^2 - 4\lambda + 3 = 0$$

has the roots $\lambda_1 = 1$ and $\lambda_2 = 3$. Hence,

$$\left(\frac{d}{dx} - 1\right)u(x) = 2e^{-x}$$

$$\left(\frac{d}{dx} - 3\right)y(x) = u(x).$$

The solution of the first equation follows

$$u' - u = 2e^{-x} \Rightarrow$$
$$\frac{d}{dx}(ue^{-x}) = 2e^{-2x} \Rightarrow$$
$$u(x) = e^x(-e^{-2x} + c_1)$$
$$= -e^{-x} + c_1 e^x.$$

Similarly, the second equation can be integrated as

$$y' - 3y = -e^{-x} + c_1 e^x \Rightarrow$$
$$\frac{d}{dx}(ye^{-3x}) = -e^{-4x} + c_1 e^{-2x}$$
$$y(x) = e^{3x}\left(\frac{1}{4}e^{-4x} - \frac{1}{2}c_1 e^{-2x} + c_2\right).$$

Hence, the general solution is

$$y(x) = c_1 e^x + c_2 e^{3x} + \frac{1}{4}e^{-x} = y_h + y_p, \quad y_p = \frac{1}{4}e^{-x}.$$

2.1.2 Second Order ODE with Variable Coefficients

We now consider few methods of solving second order ode with variable coefficients of this form

$$\boxed{y'' + P(x)y' + Q(x)y = R(x),} \tag{18}$$

using the same technique of reducing the order of the ode.

2.1.3 Homogeneous Case: $R(x) = 0$

Method of Variation of Constants: This methods applies when we already know one independent solution, for instance $y_1(x)$. Then, the other independent solution can be found by the ansatz
$$y_2(x) = c(x) y_1(x)$$
where $c(x)$ is determined by inserting this solution into the ode,
$$c'' y_1 + 2c' y_1' + y_1'' c + P(c' y_1 + c y_1') + Q c y_1 = 0.$$
This leads to a *linear* 1st order ode in $u(x) \equiv c'(x)$ which can be integrated out by the integrating factor method,
$$u' y_1(x) + [2 y_1'(x) + P(x) y_1(x)] u = 0 \Rightarrow u(x) \Rightarrow c(x) = \int dx\, u(x).$$
When we keep track of the integration constants, then we actually obtain the general solution
$$y(x) = c_1 y_1(x) + c_2 y_2(x).$$

Example 9 Let us consider this equation
$$y'' + \frac{y'}{x^2} - \frac{y}{x^3} = 0, \qquad x \neq 0$$
for which you can show that $y_1(x) = x$ is a solution. We want to find the other independent solution by the variation of constant method. Thus, we construct the ansatz solution
$$y_2(x) = c(x) y_1(x).$$
The corresponding equation for $u(x) = c'(x)$ is
$$x^2 u' + (2x + 1) u = 0$$
and can be solved by the separation of variables as follows
$$\frac{du}{u} = -\frac{2x+1}{x^2} dx \Rightarrow$$
$$\ln u(x) = -2 \ln x + \frac{1}{x}$$
$$u(x) = \frac{1}{x^2} e^{1/x},$$

up to the arbitrary integration constant which we disregard here. Integrating it once more, we find
$$\frac{dc}{dx} = \frac{1}{x^2}e^{1/x} \rightarrow c(x) = -e^{1/x},$$
up to the second integration constant. Thus, the other independent solution is
$$y_2(x) = c(x)y_1(x) = -xe^{1/x}$$
and the general solution is then
$$y(x) = c_1 x - c_2 x e^{1/x}.$$

2.1.4 Non-Homogeneous Case: $R(x) \neq 0$

Method of Factorization: This is a generic method of finding a particular solution $y_p(x)$ when we know at least one of the independent solutions, for instance $y_1(x)$. We construct an ansatz for the particular solution given by
$$y_p = v(x)y_1(x),$$
with the corresponding derivatives
$$y_p' = y_1' v + y_1 v', \qquad y_p'' = y_1'' v + 2y_1' v' + y_1 v''.$$
We determine the equation satisfied by $v(x)$ by inserting this particular solution into the non-homogeneous ode
$$v(y_1'' + Py_1' + Qy_1) + v'(2y_1' + y_1 P) + v'' y_1 = R,$$
and reduce it to a *linear first order ode* in v'
$$v'' y_1 + v'(2y_1' + y_1 P) = R.$$
We can find the solution of v' up to an arbitrary integration constant, and integrate it once more to determine $v(x)$ up to another integration constant. When we also include the contributions with the arbitrary integration constants, we then obtain the general solution
$$y(x) = c_1 y_1(x) + c_2 y_2(x) + y_p(x).$$
By neglecting the integration constant, we are hunting only for the particular solution.

Example 10 Find the particular solution of this equation

$$y'' + \frac{y'}{x^2} - \frac{y}{x^3} = \frac{5}{x^3}.$$

Solution: Let us use again $y_1(x) = x$ to construct the ansatz of the particular solution as

$$y_p = xv(x).$$

The corresponding 1st order ODE for $u(x) = v'(x)$ is then

$$xu' + \left(2 + \frac{1}{x}\right)u = 5x^3$$

or equivalently

$$u' + \left(\frac{2}{x} + \frac{1}{x^2}\right)u = \frac{5}{x^4}.$$

This can be integrated by multiplying on both sides with $e^{2\ln x - 1/x} = x^2 e^{-1/x}$, such that

$$\frac{d}{dx}\left(ux^2 e^{-1/x}\right) = \frac{5}{x^2} e^{-1/x}.$$

By integrating this up to the arbitrary constant, we find

$$ux^2 e^{-1/x} = 5e^{-1/x} \rightarrow u(x) = \frac{5}{x^2}.$$

Thus,

$$v'(x) = \frac{5}{x^2} \rightarrow v(x) = -\frac{5}{x}$$

which implies that one particular solution is

$$y_p = -5.$$

Variation of Parameters: This is another generic method of finding a particular solution $y_p(x)$ once we know the homogeneous solution

$$y_h = c_1 y_1(x) + c_2 y_2(x).$$

This is based on the following ansatz

$$\boxed{y_p(x) = f_1(x) y_1(x) + f_2(x) y_2(x).} \qquad (19)$$

By taking successive derivatives of $y_p(x)$

$$y'_p = f_1 y'_1 + f_2 y'_2 \tag{20}$$
$$y''_p = f'_1 y'_1 + f_1 y''_1 + f'_2 y'_2 + f_2 y''_2, \tag{21}$$

and inserting them into the main ode, we find that $f_1(x)$, $f_2(x)$ satisfy

$$f'_1 y'_1 + f'_2 y'_2 = R(x).$$

We combine this with the additional constraint

$$f'_1 y_1 + f'_2 y_2 = 0,$$

for arrive at this set of equations

$$\begin{cases} f'_1 y_1 + f'_2 y_2 = 0 \\ f'_1 y'_1 + f'_2 y'_2 = R(x). \end{cases}$$

This has a unique solution when the determinant—the *Wronskian*—is nonzero for any x, namely

$$W(x) \equiv \begin{vmatrix} y_1(x) & y_2(x) \\ y'_1(x) & y'_2(x) \end{vmatrix} \neq 0.$$

The Wronskian is indeed non-zero because y_1 and y_2 are independent functions. Thus, the solutions for f'_1 and f'_2 are

$$f'_1(x) = -\frac{y_2(x)}{W(x)} R(x), \quad f'_2(x) = \frac{y_1(x)}{W(x)} R(x), \tag{22}$$

which, by integration, leads to the particular solution

$$\boxed{y_p(x) = -y_1(x) \int dx \, \frac{y_2(x)}{W(x)} R(x) + y_2(x) \int dx \, \frac{y_1(x)}{W(x)} R(x).} \tag{23}$$

Example 11 Find the particular solution $y_p(x)$ for this ode

$$y'' - y = \cosh x$$

using the variation of parameters method.

Solution: The homogeneous equation reduces to

$$y'' - y = 0$$

and has the independent solutions $y_1(x) = e^x$ and $y_2(x) = e^{-x}$. We compute their associated Wronskian at arbitrary point x

$$W(x) \equiv \begin{vmatrix} e^x & e^{-x} \\ e^x & -e^{-x} \end{vmatrix} = -2 \neq 0.$$

The solutions for $f_1(x)$ and $f_2(x)$ follow from integrating their corresponding equations

$$f_1'(x) = \frac{e^{-x}}{2}\cosh(x) = \frac{e^{-x}}{2}\frac{e^x + e^{-x}}{2}$$

$$f_2'(x) = -\frac{e^x}{2}\cosh(x) = -\frac{e^x}{2}\frac{e^x + e^{-x}}{2},$$

namely,

$$f_1'(x) = \frac{1 + e^{-2x}}{4} \Rightarrow f_1(x) = \frac{1}{4}(x - \frac{1}{2}e^{-2x})$$

$$f_2'(x) = -\frac{e^{2x} + 1}{4} \Rightarrow f_1(x) = -\frac{1}{4}(x + \frac{1}{2}e^{2x}).$$

Thus, the particular solution is

$$y_p(x) = \frac{1}{4}x(e^x - e^{-x}) - \frac{1}{8}(e^x + e^{-x}).$$

Example 12 Solve the same ode

$$y'' - y = \cosh x,$$

using the factorization method instead.

Solution: We start with the ansatz for the particular solution $\tilde{y}_p(x) = c(x)e^x$ and insert it into our ode, we obtain a first order ODE for $c'(x)$,

$$c'' + 2c' = e^{-x}\frac{e^x + e^{-x}}{2}.$$

We may solve this by the integration factor method

$$\frac{d}{dx}(c'e^{2x}) = \frac{e^{2x} + 1}{2} \Rightarrow c' = \frac{1}{4} + \frac{1}{2}xe^{-2x} \Rightarrow c(x) = \frac{1}{4}x(1 - e^{-2x}) - \frac{1}{8}e^{-2x}$$

Thus, the particular solution is

$$\tilde{y}_p(x) = \frac{1}{4}x(e^x - e^{-x}) - \frac{1}{8}e^{-x}$$

Notice that the particular solution is **not unique**. However, we must always check that $y_p - \tilde{y}_p$ is a linear combinations of the independent solutions,

$$y_p - \tilde{y}_p = -\frac{1}{8}e^x.$$

2.1.5 Homogeneous Euler-Cauchy Equation

The Euler-Cauchy equation is intimately related to solving the Laplace's equation in polar coordinates, and is relevant to many physics applications in quantum and classical physics. The homogeneous Euler-Cauchy equation is given in the general form as

$$\boxed{x^2 y'' + a_1 x y' + a_0 y = 0} \qquad (24)$$

where a_1 and a_0 are constants.

Method 1: We use the ansatz that the independent solutions are power functions x^m and determine m from its characteristic equation

$$[m(m-1) + a_1 m + a_0]x^m = 0 \Rightarrow m^2 + (a_1 - 1)m + a_0 = 0, \qquad x \neq 0.$$

If $m_1 \neq m_2$, then the general solution follows as

$$y(x) = c_1 x^{m_1} + c_2 x^{m_2}, \qquad m_1 \neq m_2.$$

If $m_1 = m_2 = m$, then we obtain one independent solution $y_1 = x^m$ and can determine the other one by the variation of constant method

$$y_2(x) = c(x) y_1(x) = c(x) x^m.$$

Example 13 Find the general solution of this Euler-Cauchy equation

$$x^2 y'' - 9y = 0 \qquad (25)$$

Solution: By inserting the ansatz x^m into the ode, we find the characteristic equation for m,

$$m(m-1) - 9 = 0 \Rightarrow m_{1,2} = \frac{1}{2} \pm \frac{\sqrt{37}}{2}$$

Thus, the general solution follows by a linear superposition of the independent solutions,

$$y(x) = c_1 x^{m_1} + c_2 x^{m_2}$$

which is valid for $x \neq 0$.

Example 14 Find the general solution of this Euler-Cauchy equation

$$x^2 y'' - xy' + y = 0. \tag{26}$$

Solution: The corresponding characteristic equation for m is

$$m^2 - 2m + 1 = 0 \Rightarrow m_1 = m_2 = 1.$$

Thus, $y_1(x) = x$ and $y_2(x) = c(x)x$, where $c(x)$ is obtained by inserting $y_2(x)$ in Eq. (26):

$$xc'' + c' = 0 \Rightarrow$$
$$\frac{dc'}{c'} = -\frac{dx}{x} \Rightarrow$$
$$c' = \frac{1}{x} \rightarrow c = \ln x$$

and $y_2(x) = x \ln x$. The general solution is given as

$$y(x) = c_1 x + c_2 x \ln x, \quad x > 0$$

A similar general form holds also for $x < 0$, but the coefficient might be different

$$y(x) = \tilde{c}_1 x + \tilde{c}_2 x \ln(-x), \quad x < 0.$$

Change of Variable Method: Another approach to solving the Euler-Cauchy equation is to use the following change of variable

$$x = e^z, \quad x > 0$$
$$x = -e^z, \quad x < 0,$$

such that the Eq. (24) transforms to an ODE with *constant* coefficients

$$\frac{d^2 y}{dz^2} + (a_1 - 1)\frac{dy}{dz} + a_0 y(z) = 0. \tag{27}$$

Then, the general solution takes the form

$$y(x) = y(z) = c_1 y_1(z) + c_2 y_2(z) = c_1 y_1(\ln x) + c_2 y_2(\ln x)$$

However, we need to consider separately the two branches of solutions for $x > 0$ and $x < 0$.

Example 15 Solve Eq. (25) by the change of variable method.

Solution: By using the substitution $x = e^z$ for $x > 0$, we rewrite Eq. (25) as

$$\frac{d^2y}{dz^2} - \frac{dy}{dz} - 9y(z) = 0 \Rightarrow y(z) = c_1 e^{m_1 z} + c_2 e^{m_2 z}$$

where

$$m_{1,2} = \frac{1}{2} \pm \frac{\sqrt{37}}{2}$$

Thus, the general solution is

$$y(x) = c_1 x^{m_1} + c_2 x^{m_2}$$

which is valid for $x \neq 0$ and the coefficients c_1, c_2 could be different for the two branches of solutions ($x < 0$ and $x > 0$).

3 Lecture 15: Green Function Method

The Green function method is a versatile tool for solving linear problems in various areas of physics. Green functions are used to determine the linear response of a system at one point due to a small perturbation at another point in space and time. This method is particularly useful in cases involving localized force sources, such as the electric field from a point charge or the temperature distribution from a concentrated heat source.

In this chapter, we will introduce the Green function method for solving second-order ordinary differential equations. We will begin by discussing the Dirac delta function, which is used to represent pointwise forcing terms.

3.1 Dirac Delta Function

To prepare the ground for the Green function method, let us first introduce the Dirac delta function $\delta(x)$ and its associated calculus. The δ function is a highly-peaked distribution function defined through its integral:

$$\boxed{\int_a^b dx f(x) \delta(x - x_0) = \begin{cases} f(x_0) & \text{if } a < x_0 < b \\ 0, & \text{otherwise} \end{cases}} \tag{28}$$

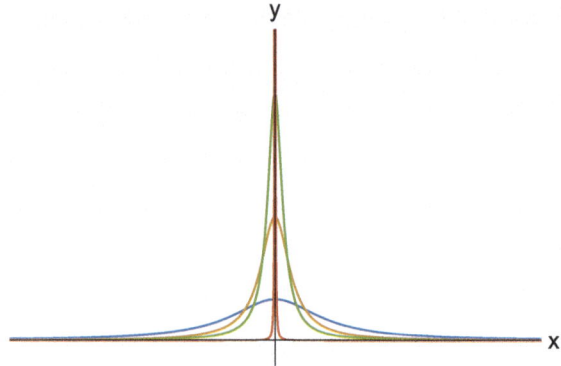

Fig. 1 Example of sequence functions approaching the Dirac delta function

The particular case of $f(x) = 1$ leads to the normalization condition of the δ distribution

$$\int_{-\infty}^{\infty} dx \delta(x) = 1. \tag{29}$$

The Dirac delta function can be defined in terms of the Dirac delta sequences of strongly-peaked functions $\{\phi_n(x)\}$ for $n = 1, 2, \cdots$, such that

$$\int_{-\infty}^{\infty} dx \delta(x) \equiv \lim_{n \to \infty} \int_{-\infty}^{\infty} dx \phi_n(x - a) = 1 \tag{30}$$

and, more generally,

$$\int_{-\infty}^{\infty} dx f(x) \delta(x - x_0) \equiv \lim_{n \to \infty} \int_{-\infty}^{\infty} dx \phi_n(x - a) f(x) = f(a). \tag{31}$$

For this definition, Eqs. (28) and (29) represent a short-hand notation of the limit integration over the Dirac delta sequences. Notice that the sequence functions ϕ_n do not have a proper limit by themselves, i.e. only the limit of their integral is well-defined and thus they are "meaningful" only inside integrals. We can see this better by considering some examples of Dirac delta sequences (Fig. 1).

Let us consider the example of a sequence function that is constant in the interval centered at origin and zero everywhere,

$$\phi_n(x) = \begin{cases} 0 & |x| \geq \frac{1}{n} \\ \frac{n}{2}, & |x| \leq \frac{1}{n} \end{cases} \tag{32}$$

These functions ϕ_n are normalized for any n,

$$\int_{-\infty}^{\infty} dx \phi_n(x) = \frac{n}{2} \int_{-1/n}^{1/n} dx = 1$$

3 Lecture 15: Green Function Method

Also,

$$\int_{-\infty}^{\infty} dx \phi_n(x) f(x) = \frac{n}{2} \int_{-1/n}^{1/n} f(x)dx = f(y), \quad -\frac{1}{n} \le y \le \frac{1}{n}$$

where we have used the **mean value theorem** to evaluate the integral. The mean value theorem states that for $f(x)$ continuous in the interval $x \in [a, b]$, there exits a value $y \in [a, b]$ such that the area under the curve equals the area of the rectangle,

$$\int_a^b dx f(x) = f(y)(b-a).$$

Thus, in the limit of $n \to \infty$, the internal of non-zero values tends becomes infinitely small such that $y \to 0$ and

$$\lim_{n \to \infty} \int_{-\infty}^{\infty} dx \phi_n(x) f(x) = f(0).$$

Other common examples of Dirac delta sequences:

$$\phi_n(x) = \frac{n}{\pi} \frac{1}{1+n^2 x^2}$$

$$\phi_n(x) = \frac{n}{\sqrt{\pi}} e^{-n^2 x^2}$$

$$\phi_n(x) = \frac{1}{n\pi} \frac{\sin^2(nx)}{x^2}.$$

Show that Eqs. 30 and 31 are satisfied by each of these delta sequences.

Heaviside Function is the step function defined as

$$H(x-a) = \begin{cases} 0 & x < a \\ 1, & x > a \end{cases},$$

and corresponds to the anti-derivative of the Dirac delta function. It is defined as the limit of the *smooth* Heaviside sequence functions

$$H(x) \equiv \lim_{n \to \infty} \psi_n(x),$$

where $\psi_n(x)$ are the anti-derivatives of the Dirac delta sequence functions,

$$\psi_n(x) = \int_{-\infty}^x \phi_n(x')dx' \Leftrightarrow \frac{d\psi_n}{dx} = \phi_n(x).$$

Dirac Delta Calculus refers to real calculus operations involving delta functions, e.g. Equations (28)–(29), with the understanding that this is a shorthand notation for the proper calculus involving the delta sequences. This also includes integration by parts. Since there exists delta sequences of differentiable functions, we perform integration by parts on the integral

$$\int_{-\infty}^{\infty} \phi_n'(x) f(x) dx = [\phi_n(x) f(x)]_{-\infty}^{\infty} - \int_{-\infty}^{\infty} \phi_n(x) f'(x) dx$$

The surface term vanishes, since each ϕ_n is a peaked function that decays fast at $\pm\infty$. Thus, taking the limit of $r \to \infty$, we obtain that

$$\lim_{n \to \infty} \int_{-\infty}^{\infty} \phi_n'(x) f(x) dx = - \lim_{n \to \infty} \int_{-\infty}^{\infty} \phi_n(x) f'(x) dx$$

In the shorthand notation, this can be written as

$$\int_{-\infty}^{\infty} \delta'(x) f(x) dx = -f'(0).$$

By a position shift to $\delta(x - a)$, we get the more general result

$$\int_{-\infty}^{\infty} \delta'(x - a) f(x) dx = -f'(a).$$

The generalization to higher-order derivatives of the delta function is that

$$\int_{-\infty}^{\infty} \delta^{(k)}(x - a) f(x) dx = (-1)^k f^{(k)}(a).$$

3.2 Linear Second Order Ode's: Green Function Method

Let us consider a non-homogeneous, linear second order ode in the normal form

$$Dy(x) = R(x),$$

where D is a *second order, linear differential* operator

$$D \equiv \frac{d^2}{dx^2} + P(x) \frac{d}{dx} + Q(x)$$

and $R(x)$ is the forcing term. The Green function $G(x, z)$ is uniquely determined by the homogeneous part of the ODE, along with **specific** boundary conditions. It

3 Lecture 15: Green Function Method

is **independent** of the forcing term $R(x)$. In principle, we could have any boundary conditions, but for now we consider either **homogeneous boundary conditions**:

$$y(a) = y(b) = 0, \tag{33}$$

or the **homogeneous initial conditions**

$$y(0) = y'(0) = 0. \tag{34}$$

The **specific** solution under boundary conditions is determined by a **convolution** integral of the Green function with the forcing term $R(x)$:

$$\boxed{y(x) = \int_a^b dz\, G(x, z) R(z), \qquad a \leq x, z \leq b.} \tag{35}$$

Similarly, the **specific** solution under initial conditions is determined by the **convolution** integral:

$$\boxed{y(x) = \int_0^\infty dz\, G(x, z) R(z), \qquad x, z \geq 0.} \tag{36}$$

The Green function is determined by the following properties:

Property 1: $G(x, z)$ satisfies the ode with a Dirac delta forcing:

$$\boxed{DG(x, z) = \delta(x - z).} \tag{37}$$

Notice that this ode is otherwise homogeneous, except for the singular point $x = z$ where ode is undetermined. Now, z is a variable for $G(x, z)$, so the singular point dependents on z, and we will see the implications of this later.

We can derive Eq. (37) by applying the differential operator D on both sides of the integral solution from Eq. 35:

$$Dy(x) = \int_a^b [DG(x, z)] R(z),$$

where z, x are both inside $[a, b]$. Using the differential equation

$$Dy(x) = R(x)$$

and writing the source term as the convolution integral with the Dirac delta function,

$$R(x) = \int_a^b \delta(x - z) R(z),$$

we arrive at

$$\int_a^b [DG(x, z) - \delta(x - z)]R(z) = 0.$$

The integral vanishes for an arbitrary forcing when the Green function satisfies the differential equation

$$DG(x, z) = \delta(x - z).$$

The point $x = z$ imposes additional restrictions on $G(x, z)$.

Property 2:

$$\boxed{G(x, z) \text{ is continuous at } x = z} \qquad (38)$$

Property 3:

$$\boxed{G(x, z) \text{ has a unit jump at } x = z} \qquad (39)$$

We want to determine $G(x, z)$ as a **continuous** function everywhere in the integration domain $[a, b]$, hence also at the singular point $x = z$. Integrating on both sides of Eq. 37 over an infinitesimal domain around the singular point $[z - \epsilon, z + \epsilon]$, we have that

$$\int_{z-\epsilon}^{z+\epsilon} dx\, DG(x, z) = \int_{z-\epsilon}^{z+\epsilon} dx\, \delta(x - z) = 1.$$

Using the definition of the differential operator, we find that

$$\int_{z-\epsilon}^{z+\epsilon} dx \frac{d^2}{dx^2} G(x, z) + \int_{z-\epsilon}^{z+\epsilon} dx\, P(x) \frac{d}{dx} G(x, z) + \int_{z-\epsilon}^{z+\epsilon} dx\, Q(x) G(x, z) = 1.$$

In the limit of $\epsilon \to 0$, this reduces to

$$\lim_{\epsilon \to 0} \left[\frac{d}{dx} G(x, z) \right]_{z-\epsilon}^{z+\epsilon} + P(z) \lim_{\epsilon \to 0} [G(x, z)]_{z-\epsilon}^{z+\epsilon} = 1.$$

The continuity assumption implies that

$$\lim_{\epsilon \to 0} [G(x, z)]_{z-\epsilon}^{z+\epsilon} = 0 \Leftrightarrow \lim_{\epsilon \to 0} G(z + \epsilon, z) = \lim_{\epsilon \to 0} G(z - \epsilon, z). \qquad (40)$$

Thus, it follows that the jump in the first derivative must be equal to 1,

$$\lim_{\epsilon \to 0} \left[\frac{d}{dx} G(x, z) \right]_{z-\epsilon}^{z+\epsilon} = 1 \qquad (41)$$

Property 4:

$$\boxed{G(x, z) \text{ satisfies the same boundary/initial conditions as } y(x)} \qquad (42)$$

The homogeneous boundary conditions given in Eq. 33 imply that $G(x, z)$ must also satisfy that

$$G(a, z) = 0, \qquad G(b, z) = 0.$$

Similarly, from the homogeneous initial conditions from Eq. 34, $G(x, z)$ must satisfy that

$$G(0, z) = 0, \qquad \frac{d}{dx} G(x, z) \Big|_{x=0} = 0.$$

These four properties are sufficient to determine uniquely the function $G(x, z)$, and thus by the **specific** solution $y(x)$ from either Eq. (35) or Eq. (36).

Example 16 Find the Green function corresponding to

$$y'' + y = R(x),$$

with the homogeneous boundary conditions

$$y(0) = y(\pi/2) = 0$$

and for an arbitrary forcing term $R(x)$.

Solution: The Green function is determined by following ode

$$\frac{d^2}{dx^2} G(x, z) + \frac{d}{dx} G(x, z) = \delta(x - z)$$

with the homogeneous boundary conditions

$$G(0, z) = G(\pi/2, z) = 0.$$

The independent solutions of the homogeneous equation $y'' + y = 0$ follow quickly by successive integration method and are given by

$$y_1(x) = e^{ix}, \qquad y_2(x) = e^{-ix}.$$

Hence, the Green's function $G(x, z)$ is given by the superposition of these independent solutions. The singular point at $x = z$ introduces a jump in the coefficients for $x < z$ and $x > z$. Thus, the two branches of solutions are

$$G(x, z) = \begin{cases} a(z)e^{ix} + b(z)e^{-ix}, & x \leq z \\ c(z)e^{ix} + d(z)e^{-ix}, & x \geq z \end{cases},$$

where $0 \leq x, z \leq \pi/2$ and the functions $a(z), b(z), c(z), d(z)$ are determined by: i) the boundary conditions, ii) the continuity of $G(x, z)$ at $x = z$ and ii) the unit jump in the derivative $\frac{d}{dx} G(x, z)$ at $x = z$.

First, we notice that the boundary point $x = 0$ is located in the branch of solutions corresponding to $0 = x < z$, while $x = \pi/2$ picks up the other solution branch for $\pi/2 = x > z$.

Boundary Conditions: $G(0, z) = G(\pi/2, z) = 0$ imply that:

$$G(0, z) = a(z) + b(z) = 0 \Rightarrow b(z) = -a(z)$$

$$G(\pi/2, z) = c(z) - d(z) = 0 \Rightarrow d(z) = c(z)$$

Thus,

$$G(x, z) = \begin{cases} a(z)(e^{ix} - e^{-ix}) = \tilde{a}(z) \sin x, & x \leq z \\ c(z)(e^{ix} + e^{-ix}) = \tilde{c}(z) \cos x, & x \geq z \end{cases}$$

Continuity at $x = z$

$$G(x \searrow z, z) = G(x \nearrow z, z),$$

implies that

$$\tilde{a}(z) \sin z = \tilde{c}(z) \cos z \Rightarrow \tilde{c}(z) = \tilde{a}(z) \frac{\sin z}{\cos z}.$$

Unit jump at $x = z$

$$\left(\frac{dG}{dx} \right)_{x \searrow z} - \left(\frac{dG}{dx} \right)_{x \nearrow z} = 1,$$

which leads to

$$-\tilde{a}(z) \cos z - \tilde{a}(z) \frac{\sin z}{\cos z} \sin z = 1 \Rightarrow \tilde{a}(z) = -\cos(z)$$

which means that $\tilde{c}(z) = -\sin(z)$. Thus, the Green function becomes

$$G(x, z) = \begin{cases} -\cos(z) \sin(x), & x \leq z \\ -\sin(z) \cos(x), & x \geq z. \end{cases}$$

The specific solution for this homogeneous boundary conditions follows by the convolution integral

$$y(x) = \int_0^{\pi/2} dz\, G(x,z) R(z)$$

$$= -\cos(x) \int_0^x dz\, \sin(z) R(z) - \sin(x) \int_x^{\pi/2} dz\, \cos(z) R(z),$$

where the integral solution depends on the form of $R(z)$.

4 Lecture 16: Green Function Method Examples

In general, non-zero boundary/initial conditions can be incorporated into the specific solution $y_h(x)$ of the homogeneous equations ($R(x) = 0$). Then, a particular solution $y_p(x)$ is determined by the sourcing term $R(x) \neq 0$ for the homogeneous boundary/initial conditions. In this lecture, we use the Green function find particular solutions y_p under homogeneous boundary/initial conditions.

4.1 Green Function Method

In summary, the Green function $G(x,z)$ for a second order **linear and non-homogeneous** ode of this form

$$Dy(x) = R(x), \qquad D \equiv \frac{d^2}{dx^2} + P(x)\frac{d}{dx} + Q(x),$$

with either *homogeneous boundary conditions*

$$y(a) = y(b) = 0,$$

or *homogeneous initial conditions*

$$y(0) = y'(0) = 0,$$

is determined by the following properties:

1. $DG(x,z) = \delta(x-z)$, thus $x = z$ is a singular point
2. $G(x,z)$ is assumed to be *continuous* at $x = z$.
3. $\frac{d}{dx} G(x,z)$ has a jump of size 1 at $x = z$.

4. $G(x, z)$ satisfies the same boundary conditions as $y(x)$.

4.1.1 Connection to Particular Solution

Let us consider the general case again

$$Dy(x) = R(x),$$

and compute the particular solution $y_p(x)$.

We write the general expression of the Green function as a linear superposition of the corresponding independent solutions $y_1(x)$ and $y_2(x)$ on each side of the singular point at $x = z$,

$$G(x, z) = \begin{cases} a(z)y_1(x) + b(z)y_2(x), & x \leq z \\ c(z)y_1(x) + d(z)y_2(x), & x \geq z \end{cases}$$

From the homogeneous initial conditions that $G(0, z) = 0$ and $\frac{d}{dx}G(x, z)|_{x=0} = 0$, it follows that

$$ay_1(0) + by_2(0) = 0$$
$$ay_1'(0) + by_2'(0) = 0.$$

Since the Wronskian

$$W(x) = \begin{vmatrix} y_1(x) & y_2(x) \\ y_1'(x) & y_2'(x) \end{vmatrix}$$

is non-zero for any x, thus also $W(0) \neq 0$, the unique solution is the trivial solution $a(z) = b(z) = 0$. Hence,

$$G(x, z) = \begin{cases} 0, & x \leq z \\ c(z)y_1(x) + d(z)y_2(x), & x \geq z. \end{cases}$$

The remaining two constants are determined from the continuity of $G(x, z)$ and the unit jump in its derivative at $x = z$:

$$cy_1(z) + dy_2(z) = 0$$
$$cy_1'(z) + dy_2'(z) = 1$$

The solution is this system of equations can be written in terms of the Wronskian $W(z)$:

$$c(z) = -\frac{y_2(z)}{W(z)}, \quad d(z) = \frac{y_1(z)}{W(z)}.$$

Thus,

$$G(x,z) = \begin{cases} 0, & x \leq z \\ -y_1(x)\frac{y_2(z)}{W(z)} + y_2(x)\frac{y_1(z)}{W(z)}, & x \geq z \end{cases}$$

The specific solution for these homogeneous conditions follows as the convolution integral

$$y(x) = \int_0^\infty dz\, G(x,z) R(z)$$
$$= -y_1(x)\int_0^x dz\frac{y_2(z)}{W(z)} R(z) + y_2(x)\int_0^x dz\frac{y_1(z)}{W(z)} R(z),$$

which is the same as the integral solution obtained using the variation of parameters method. Thus, we can determine a particular $y_p(x)$ for homogeneous conditions, as any non-zero conditions can be imposed on the solution $y_h(x) = c_1 y_1(x) + c_2 y_2(x)$ of the homogeneous equation.

4.2 Initial Value Problem

Example 17 Find the Green function for a forced harmonic oscillator given by

$$y'' + y = R(x),$$

with the homogeneous initial conditions,

$$y(0) = y'(0) = 0$$

and for an arbitrary forcing term $R(x)$.

Solution: The Green function is determined by

$$\frac{d^2}{dx^2} G(x,z) + G(x,z) = \delta(x-z)$$

with the homogeneous initial conditions

$$G(0,z) = \frac{d}{dx} G(x,z)|_{x=0} = 0.$$

For $x \neq z$, the Green function is the solution of the homogeneous equation $y'' + y = 0$, which has the independent solutions $e^{\pm ix}$. Hence, we can write the Green function as

$$G(x,z) = \begin{cases} a(z)e^{ix} + b(z)e^{-ix}, & x \leq z \\ c(z)e^{ix} + d(z)e^{-ix}, & x \geq z \end{cases},$$

where the coefficients $a(z), b(z), c(z), d(z)$ are functions of z and determined by the properties of $G(x, z)$ at the singular point $x = z$. The homogeneous initial conditions at $x = 0$ imply that

$$a(z) + b(z) = 0$$
$$a(z) - b(z) = 0$$

thus $a(z) = b(z) = 0$, and

$$G(x, z) = \begin{cases} 0, & x \leq z \\ c(z)e^{ix} + d(z)e^{-ix}, & x \geq z \end{cases}$$

The continuity of $G(x, z)$ and the unit jump in the its derive at $x = z$ lead to the following set of equations:

$$\begin{cases} c(z)e^{iz} + d(z)e^{-iz} = 0 \\ ic(z)e^{iz} - id(z)e^{-iz} = 1 \end{cases}$$

with the solution $c(z) = \frac{1}{2i}e^{-iz}$ and $d(z) = -\frac{1}{2i}e^{iz}$. Thus,

$$G(x, z) = \begin{cases} 0, & x \leq z \\ \frac{1}{2i}(e^{i(x-z)} - e^{-i(x-z)}) = \sin(x - z), & x \geq z \end{cases}$$

The corresponding specific solution under *homogeneous initial conditions* is then given by the convolution integral of the above Green function with the forcing

$$y(x) = \int_0^\infty dz\, G(x, z) R(z)$$
$$= \int_0^x dz\, \sin(x - z) R(z).$$

Example 18 Find the solution to the forced harmonic oscillator

$$y'' + y = \frac{1}{\cos x},$$

with the homogeneous initial conditions

$$y(0) = y'(0) = 0.$$

Solution: We have already derived the Green function as

$$G(x, z) = \begin{cases} 0, & x \leq z \\ \sin(x - z), & x \geq z \end{cases}$$

The specific solution is then the convolution of $G(x, z)$ with the forcing term,

$$\begin{aligned} y(x) &= \int_0^x dz \frac{\sin(x - z)}{\cos z} \\ &= \int_0^x dz \frac{\sin x \cos z - \cos x \sin z}{\cos z} \\ &= \sin x \int_0^x dz \frac{\cos z}{\cos z} - \cos x \int_0^x dz \frac{\sin z}{\cos z} \\ &= x \sin x + \cos x \ln(|\cos x|). \end{aligned}$$

Now let us apply the variation of parameters method to see that we arrive at a very similar expression for the particular solution $y_p(x)$. Let us compute the Wronskian corresponding to the independent solutions $y_1(x) = \sin x$ and $y_2(x) = \cos x$:

$$W(x) \equiv \begin{vmatrix} y_1(x) & y_2(x) \\ y_1'(x) & y_2'(x) \end{vmatrix} = \begin{vmatrix} \sin x & \cos x \\ \cos x & -\sin x \end{vmatrix} = -1$$

Thus, from the variation of parameters method, the particular solution is given by

$$\begin{aligned} y_p(x) &= -y_1(x) \int dx \frac{y_2(x)}{W(x)} R(x) + y_2(x) \int dx \frac{y_1(x)}{W(x)} R(x) \\ &= \sin x \int dx \frac{\cos x}{\cos x} - \cos x \int dx \frac{\sin x}{\cos x} \\ &= x \sin x - \cos x \ln(|\cos x|). \end{aligned}$$

4.3 Boundary Value Problem

Example 19 Find the specific solution for the forced harmonic oscillator

$$y'' + y = \frac{1}{\cos z},$$

with the homogeneous boundary conditions

$$y(0) = y(\pi/2) = 0.$$

Find the particular solution $y_p(x)$.

Solution: The corresponding Green function was derived in the previous lecture and given as

$$G(x, z) = \begin{cases} -\cos z \sin x, & x \leq z \\ -\sin z \cos x, & x \geq z \end{cases}$$

Thus, the specific solution follows as the convolution:

$$y(x) = \int_0^{\pi/2} dz\, G(x, z) \frac{1}{\cos z}$$

$$= -\sin x \int_x^{\pi/2} dz \frac{\cos z}{\cos z} - \cos x \int_0^x dz \frac{\sin z}{\cos z}$$

$$= -\sin x \int_x^{\pi/2} dz \frac{\cos z}{\cos z} - \cos x \int_0^x dz \frac{\sin z}{\cos z}$$

$$= -\left(\frac{\pi}{2} - x\right) \sin x + \cos x \ln(|\cos x|).$$

Notice that this is also a particular solution.

Example 20 Find the solution for the damped, harmonic oscillator

$$y'' + 2y' + y = f(t)$$

where

$$f(t) = \begin{cases} 1, & 0 < t < a \\ 0, & t > 0 \end{cases}$$

with the initial condition $y(0) = y'(0) = 0$.

Solution: The corresponding differential operator is

$$D = \frac{d^2}{dt^2} + 2\frac{d}{dt} + 1.$$

The independent solutions can be determined by the successive integration method and equal to

$$y_1(t) = e^{-t}, \qquad y_2(t) = te^{-t}$$

The Green's function for the homogeneous initial condition is then

$$G(t, t') = \begin{cases} 0, & t \leq t' \\ c(t')e^{-t} + d(t')te^{-t}, & t \geq t' \end{cases}$$

From the properties of $G(t, t')$ at $t = t'$, it follows that

$$\begin{cases} ce^{-t'} + dt'e^{-t'} = 0 \\ -ce^{-t'} + d(1-t')e^{-t'} = 1 \end{cases}$$

with the solutions $c = -t'e^{t'}$ and $d = e^{t'}$. Thus,

$$G(t,t') = \begin{cases} 0, & t \leq t' \\ (t-t')e^{-(t-t')}, & t \geq t' \end{cases}$$

The specific solution follows from evaluating the convolution integral as

$$\begin{aligned} y(t) &= \int_0^t dt'(t-t')e^{-(t-t')} f(t') \\ &= \begin{cases} \int_0^t dt'(t-t')e^{-(t-t')}, & 0 < t < a \\ \int_0^a dt'(t-t')e^{-(t-t')}, & t > a \end{cases} \\ &= \begin{cases} e^{-t}(t-1) + 1, & 0 < t < a \\ (1-t)e^{-t} - (1-t+a)e^{-(t-a)}, & t > a \end{cases} \end{aligned}$$

5 Lecture 17: Power Series Method

Some odes might have solutions that do not have a closed form in terms of elementary functions. The power series method allows us to generate power series solutions to odes. Let us consider the normal form of a second order *homogeneous and linear* ode given by

$$y'' + p(x)y' + q(x)y = 0, \tag{43}$$

where $p(x)$ and $q(x)$ are some elementary functions of x. When $p(x)$ and $q(x)$ can be Taylor expanded around some expansion point x_0, the solution of Eq. 43 can be written as a **power series**

$$\boxed{y(x) = \sum_{n=0}^{\infty} a_n (x - x_0)^n} \tag{44}$$

This method can be generalized to series expansions around an arbitrary x_0 not necessarily a regular point, in which case the power series may have have *non-integer* exponents. This generalised power series is also called the **Frobenius series** and is given as

$$\boxed{y(x) = x^s \sum_{n=0}^{\infty} a_n (x - x_0)^n, \qquad s \text{ is real or complex}} \qquad (45)$$

The ode has a valid solution in the domain of x where the power series are convergent. In this lecture, we focus on generating the independent solutions using the power series method.

5.1 Power Series Method

This method applies when $p(x)$ and $q(x)$ are regular functions at x_0. This means that they can be Taylor expanded around x_0. Then, the solution can be represented as a power series

$$y(x) = \sum_{n=0}^{\infty} a_n (x - x_0)^n = a_0 + a_1(x - x_0) + a_2(x - x_0)^2 + \cdots$$

with coefficients a_n determined by Eq. 43. The first and second derivatives of the solution $y(x)$ follow straightforwardly through term-by-term differentiations and are also power series,

$$y'(x) = \sum_{n=1}^{\infty} n a_n (x - x_0)^{n-1} = a_1 + 2a_2(x - x_0) + 3a_3(x - x_0)^2 + \cdots$$

$$y''(x) = \sum_{n=2}^{\infty} n(n-1) a_n (x - x_0)^{n-2} = 2a_2 + 6a_3(x - x_0) + \cdots .$$

The gist of the method is that we insert these power series into Eq. 43, collect terms with equal powers of $(x - x_0)$ and set their corresponding coefficients to zero order by other. In this way, we generate a set of recursive equations for a_n's which can be solved by the iterative method.

We will apply this technique for the **Legendre equation** and find that its independent solutions reduce to the finite power series called the **Legendre polynomials**.

5.1.1 Legendre Equation

The Legendre equation is an important ode that arises in many physics problems that involve the spherical symmetry and Laplace operator ($\nabla^2 = \partial_x^2 + \partial_y^2 + \partial_z^2$). The Legendre equation is defined as

$$(1 - x^2) y'' - 2x y' + l(l+1) y = 0, \qquad l = 0, 1, 2 \ldots . \qquad (46)$$

5 Lecture 17: Power Series Method

Relating it to the canonical form, we identify the coefficients $p(x)$ and $q(x)$ as

$$p(x) = -\frac{2x}{1-x^2}, \quad q(x) = \frac{l(l+1)}{1-x^2}, \quad \text{for } x^2 \neq 1.$$

Notice that $p(x)$ and $q(x)$ are both analytic at $x_0 = 0$ which indicates that we can use the power series expansion around this point to write the general solution as

$$y(x) = \sum_{n=0}^{\infty} a_n x^n = a_0 + a_1 x + a_2 x^2 + \cdots.$$

The convergence domain of this power series is the same as the convergence domain of the Taylor series of $p(x)$ and $q(x)$ at $x_0 = 0$. It is determined by the geometric series

$$\frac{1}{1-x^2} = \sum_{n}^{\infty} x^{2n}$$

which is convergent for $|x| < 1$.

We differentiate the general solution term-by-term to obtain the Taylor series for its first two derivatives,

$$y'(x) = \sum_{n=1}^{\infty} n a_n x^{n-1}, \quad y''(x) = \sum_{n=2}^{\infty} n(n-1) a_n x^{n-2}.$$

Next, we insert these series expansions into Eq. (46)

$$(1-x^2) \sum_{n=2}^{\infty} n(n-1) a_n x^{n-2} - 2x \sum_{n=1}^{\infty} n a_n x^{n-1} + l(l+1) \sum_{n=0}^{\infty} a_n x^n = 0$$

By absorbing the x-dependent terms coefficients inside the series

$$\sum_{n=0}^{\infty} (n+2)(n+1) a_{n+2} x^n - \sum_{n=2}^{\infty} n(n-1) a_n x^n$$

$$- 2 \sum_{n=1}^{\infty} n a_n x^n + l(l+1) \sum_{n=0}^{\infty} a_n x^n = 0$$

and rearranging terms, we arrive at

$$\sum_{n=0}^{\infty} [(n+2)(n+1) a_{n+2} + l(l+1) a_n] x^n - \sum_{n=2}^{\infty} n(n-1) a_n x^n - 2 \sum_{n=1}^{\infty} n a_n x^n = 0.$$

Equivalently, we can rewrite this as

$$2a_2 + l(l+1)a_0 + [6a_3 + (l(l+1) - 2)a_1]x$$
$$+ \sum_{n=2}^{\infty} \{(n+2)(n+1)a_{n+2} + [l(l+1) - n(n-1) - 2n]a_n\}x^n = 0$$

For this power series to equal zero for any x, the coefficients in front of each power of x must vanish. This leads to recursion formulas for the coefficients a_n, namely

$$2a_2 + l(l+1)a_0 = 0$$
$$6a_3 + (l^2 + l - 2)a_1 = 0$$
$$(n+2)(n+1)a_{n+2} + [l^2 + l - n^2 - n]a_n = 0, \quad n \geq 2.$$

which imply that

$$a_2 = -\frac{l(l+1)}{2}a_0$$
$$a_3 = -\frac{(l-1)(l+2)}{6}a_1$$
$$a_{n+2} = -\frac{(l-n)(l+n+1)}{(n+2)(n+1)}a_n, \quad n \geq 2$$

From the recursive relation, we notice that all coefficients of even powers x^{2k} are generated by a_0, namely

$$a_2 = -\frac{l(l+1)}{2}a_0$$
$$a_4 = -\frac{(l-2)(l+3)}{3 \cdot 4}a_2 = (-1)^2 \frac{l(l-2) \cdot (l+1)(l+3)}{4!}a_0$$
$$\ldots$$
$$a_{2n} = (-1)^n \frac{l(l-2) \cdots (l-2n+2) \cdot (l+1)(l+3) \cdots (l+2n-1)}{(2n)!}a_0$$

while all coefficients of odd powers x^{2k+1} are generated by a_1,

$$a_3 = -\frac{(l-1)(l+2)}{2 \cdot 3}a_1$$
$$a_5 = -\frac{(l-3)(l+4)}{4 \cdot 5}a_1 = (-1)^2 \frac{(l-1)(l-3) \cdot (l+2)(l+4)}{5!}a_1$$
$$\ldots$$
$$a_{2n+1} = (-1)^n \frac{(l-1)(l-3) \cdots (l-2n+1) \cdot (l+2)(l+4) \cdots (l+2n)}{(2n+1)!}a_1$$

We notice that a_0 and a_1 are two arbitrary constants that factor out in front of the corresponding power series to become the constants in front of the independent

5 Lecture 17: Power Series Method

solutions. Thus, the general solution of Eq. (46) can be expressed as

$$y(x) = a_0 y_1(x) + a_1 y_2(x),$$

where the independent solutions y_1 and y_2 are the power series

$$y_1(x) = \sum_{n=0}^{\infty} (-1)^n \frac{l(l-2)\cdots(l-2n+2)\cdot(l+1)(l+3)\cdots(l+2n-1)}{(2n)!} x^{2n}$$

$$y_2(x) = \sum_{n=0}^{\infty} (-1)^n \frac{(l-1)(l-3)\cdots(l-2n+1)\cdot(l+2)(l+4)\cdots(l+2n)}{(2n+1)!} x^{2n+1}$$

Convergence Interval: As expected from the structure of $p(x)$ and $q(x)$, the series should be convergent for $|x| < 1$, and we can double-check it using the ratio test.

For $y_1(x)$:

$$\rho_1 = \lim_{n \to \infty} \frac{(l-2n-4)\cdot(l-2n+1)}{(2n+1)(2n+2)} |x|^2 < 1 \Rightarrow |x|^2 < 1.$$

For $y_2(x)$:

$$\rho_2 = \lim_{n \to \infty} \frac{(l-2n-1)\cdot(2n+4)}{(2n+3)(2n+3)} |x|^2 < 1 \Rightarrow |x|^2 < 1.$$

Important Point: The convergence at the end points $x = \pm 1$ depends on specific values of the parameter l. It can be shown that for a positive integer l, only one of the independent solutions is convergent at $x = \pm 1$ and becomes a polynomial of degree l which is called the **Legendre polynomial**. We distinguish two cases, namely when l is even or odd.

For $l = 2m$: Let us take the lowest value of the parameter $l = 0$. From the recursive relations in Eq. (47), we see that all coefficients depend on l thus are also zero, expect for a_0.

Now let us consider the next order $l = 2$, for which only a_0 and

$$a_2 = -3a_0 \neq 0$$

while all the other coefficients becomes zero because of the $(l-2)$ term that vanishes.

Perhaps we start to see the pattern that for an even parameter $l \equiv 2m$ only the first coefficients up to a_{2m} are non-zero and all other vanish. Hence, the infinite series reduces to a polynomial of degree $l = 2m$.

In general, the products of l's in Eq. (47) can be written in terms of factorials, namely

$$l(l-2)\cdots(l-2n+2) = 2m(2m-2)\cdots(2m-2n+2)$$
$$= \frac{2^n m!}{(m-n)!}$$
$$(l+1)(l+3)\cdots(l+2n-1) = (2m+1)(2m+3)\cdots(2m+2n-1)$$
$$= \frac{(2m+2n)!m!}{2^n(2m)!(m+n)!}$$

such that the coefficients of even powers are

$$a_{2n} = (-1)^n \frac{(m!)^2}{(2m)!} \frac{(2m+2n)!}{(m-n)!(m+n)!(2n)!} a_0 \qquad (47)$$

which are non-zero for $n < m$ and zero otherwise. This means that the independent solution $y_1(x)$ reduces to a finite sum given by a polynomial of degree $2m$, namely

$$y_1(x) \equiv Q_{2m}(x) = \frac{(m!)^2}{(2m)!} \sum_{n=0}^{m} (-1)^n \frac{(2m+2n)!}{(m+n)!(m-n)!(2n)!} x^{2n}$$

which determines the Legendre polynomial $P_n(x)$ up to some rescaling constant fixed by the boundary condition. Thus, there is a parametric family of Legendre equations for different values of even values of l parameter $l = 0, 2, \ldots, 2m$ where $y_1(x)$ is given by the Legendre polynomial of degree l. For instance, the polynomials of lowest degrees are

$$P_0(x) = Q_0(x) = 1$$
$$P_2(x) = -\frac{1}{2} Q_2(x) = \frac{1}{2}(3x^2 - 1)$$

Note that the other independent solution $y_2(x)$ remains an infinite series with odd powers of x, because the coefficients of x^{2n+1} are never zeroes. Therefore $y_2(x)$ is divergent at $x = \pm 1$. For example, let us consider the case with $l = 0$ and see that the recursive relations for a_{2n+1} give non-zero a_{2n+1} for any n:

$$a_3 = -\frac{(l-1)(l+2)}{2 \cdot 3} a_1 = \frac{1}{3} a_1$$
$$a_5 = -\frac{(l-3)(l+4)}{4 \cdot 5} a_1 = \frac{1}{5} a_1$$
$$\ldots$$
$$a_{2n+1} = -\frac{(l-2n+1)(l+2n)}{2n \cdot (2n+1)} a_{2n-1} = \frac{1}{2n+1} a_1$$

Thus,

$$y_2(x) = \sum_{n=0}^{\infty} \frac{1}{2n+1} x^{2n+1}$$

which diverges at $x = \pm 1$.

For $l = 2m + 1$: Now, let us consider the lowest odd parameter $l \equiv 1$. From the recursive relations in Eq. (47), it follows that the corresponding coefficients in $y_2(x)$ are

$$a_{2n+1} = 0, \quad \text{for } n \geq 1$$

hence the second independent solution is simply

$$y_2(x) = x$$

On the other hand, the coefficients of even exponents in $y_1(x)$ are all non-zero

$$a_2 = -\frac{l(l+1)}{2} a_0 = -a_0$$

$$a_4 = -\frac{(l-2)(l+3)}{3 \cdot 4} a_2 = -\frac{1}{3} a_0$$

$$\dots$$

$$a_{2n} = -\frac{(l-2n+2)(l+2n-1)}{(2n)(2n-1)} a_{2n-2} = \frac{(2n-3)}{(2n-1)} a_{2n-2} = -\frac{1}{(2n-1)} a_0$$

and

$$y_1(x) = -\sum_{n=0}^{\infty} \frac{1}{2n-1} x^{2n}$$

and diverges at $x = \pm 1$.

We can generalize this to any odd parameter $l \equiv 2m + 1$ and find that $y_1(x)$ remains an infinite power series, while $y_2(x)$ reduces to the polynomial of degree $2m + 1$, $Q_{2m+1}(x)$, given as

$$\boxed{y_2(x) \equiv Q_{2m+1} = \frac{(m!)^2}{(2m+1)!} \sum_{n=0}^{m} (-1)^n \frac{(2m+2n+1)!}{(m+n)!(m-n)!(2n+1)!} x^{2n+1}}$$

The first P_n polynomials are

$$P_1(x) = Q_1(x) = x$$

$$P_3(x) = -\frac{3}{2} Q_2(x) = \frac{1}{2}(5x^3 - 3x)$$

Fig. 2 Example of Legendre polynomials

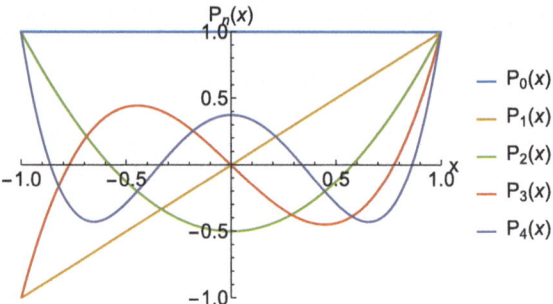

Properties of Legendre Polynomials $P_n(x)$: By setting the boundary condition so that $y(x) = 1$ when $x = 1$, the coefficients a_0 and a_1 are fixed such that the *specific solution* of the Legendre equation is the **Legendre polynomial of order** l $P_l(x)$. It is determined by the polynomial $Q_l(x)$ with the appropriate rescaling x. Figure 2 shows the dependence of the Legendre polynomials on x for $l = 0, 1, 2, 3, 4$.

The Legendre polynomials $P_n(x)$ satisfy these properties:

- $P_n(1) = 1$, for any $n \geq 0$ positive integer.
- $P_n(x) = (-1)^n P_n(-x)$, for any $n \geq 0$ positive integer.

- **Legendre polynomials are orthogonal functions**

$$\int_{-1}^{1} dx\, P_n(x) P_m(x) = \begin{cases} 0 & \text{if } m \neq n \\ \frac{2}{2n+1} & \text{if } m = n \end{cases}$$

- **Rodrigues' formula**

$$\boxed{P_n(x) = \frac{1}{2^n n!} \frac{d^n}{dx^n}(x^2 - 1)^n}$$

- **Recursion relation**

$$\boxed{n P_n(x) = (2n - 1) x P_{n-1}(x) - (n - 1) P_{n-2}(x)}$$

Generating Function for $P_n(x)$: The generating function for $P_n(x)$ is given as

$$\boxed{\Phi(x, h) = (1 - 2xh + h^2)^{-1/2}, \quad |h| < 1,}$$

such that $P_n(x)$ are given as the coefficients of the power series of $\Phi(x, h)$ with respect to h:

$$\boxed{\Phi(x, h) = \sum_{n=0}^{\infty} P_n(x) h^n}$$

5 Lecture 17: Power Series Method

For $x = 1$, the generating function reduces to

$$\Phi(1, h) = \frac{1}{1-h} = \sum_{n=0}^{\infty} h^n \Leftrightarrow P_n(1) = 1.$$

Example 21 (*Application to gravitational potential*) The generating function $\Phi(x, h)$ is useful in expanding the potential energy associated with any inverse square force. As an example, let us consider the gravitational potential between two pointwise mass objects separated by a distance $d = |\vec{R} - \vec{r}|$, where $\vec{R}$ and $\vec{r}$ are the position vectors of the mass objects relative to the origin of a 3D coordinate system. Since the gravitational force is a gradient force, $\vec{F} = -\nabla V$, and decays with the distance as $1/d^2$, it follows that the gravitational potential goes as $1/d$

$$V(d) = \frac{K}{d}, \quad K \text{ appropriate constant.}$$

We denote $R = |\vec{R}|$ and $r = |\vec{r}|$. The distance can be rewritten in terms of r/R as

$$d = \sqrt{R^2 - 2\vec{R}\cdot\vec{r} + r^2} = \sqrt{R^2 - 2Rr\cos(\theta) + r^2} = R\sqrt{1 - 2\frac{r}{R}\cos(\theta) + \frac{r^2}{R^2}}$$

We consider the case where one object is much closer to the origin than the other, $r < R$, and use the change of variables

$$h = \frac{r}{R}, \quad x = \cos\theta$$

such that the gravitational potential is written in terms of the generating function $\Phi(x, h)$ as

$$V(x, h) = \frac{K}{R}\Phi(x, h) = \frac{K}{R}\sum_{n=0}^{\infty} h^n P_n(x)$$

$$= K \sum_{n=0}^{\infty} \frac{r^n}{R^{n+1}} P_n(\cos\theta).$$

Let us consider a discrete configuration of mass points at positions $\vec{r}_i$ and angles θ_i from the position $\vec{R}$ of a test point where we compute the gravitational potential and the gravitational force induced by the other points. The potential at R is a superposition of the potentials induced by each mass point. In the assumption that $R > r_i$, it becomes

$$V(x, h) = \sum_i K \sum_{n=0}^{\infty} \frac{r_i^n}{R^{n+1}} P_n(\cos\theta_i).$$

In the limit of a continuous mass distribution, the sum $\sum_i$ over all point masses becomes a volume integral over the mass density

$$V = K \iiint d\vec{r} \rho(r) \sum_{n=0}^{\infty} \frac{r^n}{R^{n+1}} P_n(\cos\theta),$$

where θ is the angle between $\vec{R}$ and $\vec{r}$. When $R \gg r$, the dominant term of this expansion corresponds to $n = 0$ and the spherical approximation of the potential,

$$V_0 \approx \frac{K}{R} \iiint d\vec{r} \rho(r) = M \frac{K}{R}$$

This is effectively the gravitational potential induced by the mass M localized at the origin or when the mass is distributed spherically symmetric around the origin. Any deviation from the spherical symmetry in the mass distribution will pick up more terms in the expansion, and the potential will depend on higher order Legendre polynomials.

This has important applications to **satellite gravimetry**, when satellites orbit the Earth and measure the local gravitational field at a given orbital radius R (hence $R > r$). Since the Earth is not a perfect sphere, higher-order Legendre polynomials are used in these calculations. This approach is not limited to Earth; it can be applied to any celestial body that satellites may orbit around, allowing us to study the gravitational field of various objects in space.

6 Lecture 18: Frobenius Method

We consider a generic homogeneous and linear second order ode on the form

$$y'' + p(x) y' + q(x) y = 0, \tag{48}$$

where $p(x)$ and $q(x)$ are elementary functions of x.

In this lecture, we focus on the Frobenius series expansion which can be applied in general when $p(x)$ and $q(x)$ may not be analytic at the expansion point x_0.

Let us consider the ode of this particular form,

$$y'' + \frac{b(x)}{x} y' + \frac{c(x)}{x^2} y = 0, \tag{49}$$

such that $b(x)$ and $c(x)$ **are** analytic at $x_0 = 0$. An independent solution can be represented as a *generalized power series* also known as **Frobenius series**

6 Lecture 18: Frobenius Method

$$y(x) = x^s \sum_{n=0}^{\infty} a_n x^n, \qquad \text{where } s \text{ is real or complex}$$

We notice that power series are special cases of the Frobenius series when $s = 0$. The generic steps are the same as for the power series method. Let us Taylor expand the coefficients $b(x)$ and $c(x)$ around $x_0 = 0$,

$$b(x) = \sum_{m=0} b_m x^m, \qquad c(x) = \sum_{m=0} c_m x^m.$$

The derivative of the ansatz solution in the Frobenius form are

$$y(x) = \sum_{n=0} a_n x^{s+n}$$

$$xy'(x) = \sum_{n=0} (s+n) a_n x^{s+n}$$

$$x^2 y''(x) = \sum_{n=0} (s+n)(s+n-1) a_n x^{s+n}.$$

We insert these expansions into Eq. (49), and find that

$$\sum_{n=0} \left[(s+n-1)(s+n) a_n x^n + \sum_{m=0} [(s+n) b_m + c_m] a_n x^{n+m} \right] = 0. \qquad (50)$$

This implies that the coefficients in front of each power of x must vanish one by one. The zeroth order term leads to

$$a_0 [s(s-1) + s b_0 + c_0] = 0. \qquad (51)$$

Since by construction, $a_0 \neq 0$, it follows that s is a solution of the quadratic equation also known as the **indicial equation**,

$$s^2 + s(b_0 - 1) + c_0 = 0. \qquad (52)$$

For distinct roots, $s_1 \neq s_2$, the two Frobenius series are both solutions of the ode. However, they might not be *independent* solutions. The higher order terms in Eq. (50) lead to recursive relations for the coefficients a_n.

Example 22 Let us consider this ode

$$xy'' + y = 0$$

where $b(x) = 0$ and $c(x) = x$. Thus, $x_0 = 0$ is a singular point. The recurrence relations for a_n follow from Eq. (50) for powers larger than x^s. Adapted to this example, Eq. (50) reads as

$$\sum_{n=0} \left[(s+n-1)(s+n)a_n x^n + a_n x^{n+1}\right] = 0 \quad (53)$$

which is equivalent to

$$(s-1)sa_0 + [s(s+1)a_1 + a_0]x + [(s+1)(s+2)a_2 + a_1]x^2 + \cdots = 0 \quad (54)$$

Thus, the zeroth order term $n = 0$ corresponds to the indicial equation

$$s(s-1) = 0 \Rightarrow s_1 = 1, \quad s_2 = 0.$$

The recursive relations for a_n's are obtained by setting the rest of the coefficients to zero,

$$(s+n-1)(s+n)a_n + a_{n-1} = 0, \quad n \geq 1. \quad (55)$$

From this relation, we see that the coefficients in the Frobenius series are determined by the values of s.

Let us compute the coefficients a_n corresponding to $\underline{s_1 = 1}$,

$$a_n = -\frac{1}{n(n+1)} a_{n-1}, \quad n \geq 1.$$

This can be solved iteratively

$$a_1 = -\frac{1}{2} a_0 \quad (56)$$

$$a_2 = -\frac{1}{2 \cdot 3} a_1 = (-1)^2 \frac{1}{2^2 \cdot 3} a_0 \quad (57)$$

$$a_3 = -\frac{1}{3 \cdot 4} a_2 = (-1)^3 \frac{1}{2^2 \cdot 3^2 \cdot 4} a_0 \quad (58)$$

$$a_4 = -\frac{1}{4 \cdot 5} a_3 = (-1)^4 \frac{1}{2^2 \cdot 3^2 \cdot 4^2 \cdot 5} a_0 \quad (59)$$

and, in general,

$$a_n = (-1)^n \frac{1}{n!(n+1)!} a_0, \quad n \geq 0$$

where a_0 is arbitrary. Henceforth, one Frobenius series is given by

$$S_1(x) = \sum_{n=0}^{\infty} (-1)^n \frac{1}{n!(n+1)!} x^{n+1}.$$

For $s_2 = 0$, the corresponding recursive relation is

$$a_n = -\frac{1}{n(n-1)}a_{n-1}, \quad n \geq 2.$$

Solving it for the first few terms

$$a_2 = -\frac{1}{2}a_1$$
$$a_3 = -\frac{1}{3\cdot 2}a_2 = (-1)^2 \frac{1}{3\cdot 2^2}a_1$$
$$a_4 = -\frac{1}{4\cdot 3}a_3 = (-1)^3 \frac{1}{4\cdot 2^2 \cdot 3^2}a_1$$
$$a_5 = -\frac{1}{5\cdot 4}a_4 = (-1)^4 \frac{1}{5\cdot 2^2 \cdot 3^2 \cdot, 4^2}a_1$$

we notice that it generalizes to

$$a_n = (-1)^{n-1}\frac{1}{n!(n-1)!}a_1, \quad n \geq 1$$

where a_1 is arbitrary. Hence, the second Frobenius series is also a regular power series

$$S_2(x) = \sum_{n=1}^{\infty}(-1)^{n-1}\frac{1}{n!(n-1)!}x^n.$$

In fact, by redefining the summation index $k = n - 1$, we notice that the two series are the same

$$S_2(x) = \sum_{k=0}^{\infty}(-1)^k \frac{1}{(k+1)!k!}x^{k+1} = S_1(x).$$

For this equation, the Frobenius method gives us only one of the two independent solutions which can be expressed in terms of the **Bessel function of first kind and first order**

$$S_1(x) = \sqrt{x}J_1(2\sqrt{x}).$$

The Bessel function of first kind and of order p (integer) is an infinite power series given by

$$J_p(x) = \sum_{n=0}^{\infty}(-1)^n \frac{1}{n!(n+p)!}\left(\frac{x}{2}\right)^{2n+p}.$$

Bessel Functions J_p: The Frobenius method provides us with special functions as solutions to important ode's in physics. These are denoted in a short hand notation by simple letters as "elementary" functions, but are power series sometime with integral representations. The Frobenius method allows us to derive them and study their properties. The **Bessel equation** arises in many problems with cylindrical coordinates and has the normal form

$$x^2 y'' + xy' + (x^2 - p^2)y = 0,$$

with p being an *integer parameter* which determines the **order** of the Bessel function of the first kind $J_p(x)$. The formalism can be generalized to non-integer p's. In applying the Frobenius method, it is easier to work with an equivalent form of the Bessel equation written as

$$x(xy')' + (x^2 - p^2)y = 0 \tag{60}$$

We apply the Frobenius method to find an independent solution in the form

$$y(x) = \sum_{n=0}^{\infty} a_n x^{n+s}$$

Thus,

$$xy'(x) = \sum_{n=0}^{\infty} (n+s) a_n x^{n+s}$$

and

$$x(xy'(x))' = \sum_{n=0}^{\infty} (n+s)^2 a_n x^{n+s}.$$

Inserting these series into the Eq. 60, we have then

$$\sum_{n=0}^{\infty} [(n+s)^2 - p^2] a_n x^n + \sum_{n=0}^{\infty} a_n x^{n+2} = 0$$

By redefining the summation index in the second series, $n \to n+2$, we have

$$\sum_{n=0}^{\infty} [(n+s)^2 - p^2] a_n x^n + \sum_{n=2}^{\infty} a_{n-2} x^n = 0$$

The zero coefficient in this series leads to the indicial equation

$$[s^2 - p^2] a_0 = 0$$

which implies that $s_1 = p$ and $s_2 = -p$. The next order coefficient ($n = 1$) leads to the following relation

$$[(s+1)^2 - p^2] a_1 = 0$$

from which it follows that $a_1 = 0$. For the higher order terms, we have the following recursive relations

$$a_n = -\frac{1}{(s+n)^2 - p^2} a_{n-2}, \quad n \geq 2.$$

For $s_1 = p$, this leads to

$$a_n = -\frac{1}{(p+n)^2 - p^2} a_{n-2} = -\frac{1}{n(n+2p)} a_{n-2}.$$

Since $a_1 = 0$, all the odd terms are also zero $a_{2n+1} = 0$. The even terms are given by

$$a_{2n} = -\frac{1}{2^2 n(n+p)} a_{2n-2}.$$

Solving iteratively for each term, we find that

$$a_2 = -\frac{1}{2^2(p+1)} a_0$$

$$a_4 = -\frac{1}{2^2 2(p+2)} a_2 = (-1)^2 \frac{1}{2^4 2(p+1)(p+2)} a_0$$

$$a_4 = (-1)^2 \frac{p!}{2^4 2!(p+2)!} a_0$$

$$a_6 = -\frac{1}{2^2 3(p+3)} a_4 = (-1)^3 \frac{p!}{2^6 3!(p+3)!} a_0$$

and, in general,

$$a_{2n} = (-1)^n \frac{1}{2^{2n} n!} \frac{p!}{(p+n)!} a_0$$

where a_0 is arbitrary. The factorial $p!$ is also a constant that can be absorbed into a_0. Thus, the Frobenius series is a series with only even coefficients and generates the **Bessel function $J_p(x)$ of order p**

$$\boxed{J_p(x) = \sum_{n=0}^{\infty} (-1)^n \frac{1}{n!(p+n)!} \left(\frac{x}{2}\right)^{2n+p}}$$

where we have added 2^{-p} as part of the series so that we can have $x/2$ raised to $2n + p$.

This Bessel function is an independent solution of the Bessel equation. Similarly, we can solve the recursive relations for the coefficients corresponding to $s_2 = -p$ and find that the coefficients in front of the odd powers also vanish by the same argument, while the coefficients in front of the even powers are generated by the corresponding recursive relation given by

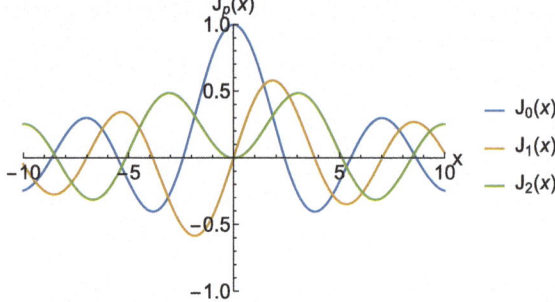

Fig. 3 Bessel functions $J_0(x)$, $J_1(x)$, $J_2(x)$ as functions of x

$$a_{2n+2} = \begin{cases} -\dfrac{1}{2^2(n+1)(n-p+1)} a_{2n}, & n \geq p \\ 0, & n < p \end{cases}$$

so that the arbitrary coefficient is a_{2p}. Let us take $p = 1$ to check this. From the main recursive relation, we see that

$$a_{2n} 2^2 n(n-1) = -a_{2n-2}.$$

For $n = 1$, it means that $a_0 = 0$, while the rest of the coefficients are generated by a_2, as

$$a_4 = -\frac{1}{2^2 2} a_2$$

$$a_6 = -\frac{1}{2^2 3 \cdot 2} a_4 = -2^2(-1)^3 \frac{1}{2^6 \cdot (3!) \cdot 2} a_2$$

$$a_8 = -\frac{1}{2^2 4 \cdot 3} a_6 = -2^2(-1)^4 \frac{1}{2^8 4! 3!} a_2$$

$$\cdots$$

$$a_{2n} = -\frac{1}{2^2 n \cdot (n-1)} a_{2n-2}$$

$$a_{2n} = -2^2(-1)^n \frac{1}{2^{2n} n!(n-1)!} a_2.$$

Since a_2 is arbitrary, we can absorb the constant coefficient -2^2 into it. The corresponding solution is then the **Bessel function** $J_{-1}(x)$ **of order** -1,

$$J_{-1}(x) = \sum_{n=1}^{\infty} (-1)^n \frac{1}{n!(n-1)!} \left(\frac{x}{2}\right)^{2n-1}.$$

By substituting $s = -1$ with $s = -p$, we obtain the **Bessel function** $J_{-p}(x)$ **of order** $-p$ as

6 Lecture 18: Frobenius Method

$$J_{-p}(x) = \sum_{n=p}^{\infty}(-1)^n \frac{1}{n!(n-p)!}\left(\frac{x}{2}\right)^{2n-p}$$

By relabeling the summation index $k = n - p$, we notice that $J_{-p}(x)$ maps into $J_p(x)$ up to a sign difference

$$J_{-p}(x) = \sum_{k=0}^{\infty}(-1)^{k+p}\frac{1}{k!(k+p)!}\left(\frac{x}{2}\right)^{2k+p} \tag{61}$$

$$= (-1)^p J_p(x). \tag{62}$$

Hence, whilst $J_{-p}(x)$ is a solution of the Bessel equation, it is not independent of $J_p(x)$! So, when p is an integer, the two Bessel functions are determined from one other by this relation

$$J_{-p}(x) = (-1)^p J_p(x).$$

However, when p is a non-integer, then the two Bessel function are in fact independent of each other.

To get an intuition on the dependence of the Bessel functions on x, Fig. 3 shows the plots of first few orders $J_p(x)$:

$$J_0(x) = \sum_{n=0}^{\infty}(-1)^n \frac{1}{(n!)^2}\left(\frac{x}{2}\right)^{2n} \tag{63}$$

$$J_1(x) = \sum_{n=0}^{\infty}(-1)^n \frac{1}{n!(n+1)!}\left(\frac{x}{2}\right)^{2n+1} \tag{64}$$

$$J_2(x) = \sum_{n=0}^{\infty}(-1)^n \frac{1}{n!(n+2)!}\left(\frac{x}{2}\right)^{2n+2}. \tag{65}$$

The functions oscillate with an amplitude that is decreases with increasing $|x|$. Also, the functions are well-defined for x real, which means that the infinite series are convergent on the real axis.

Other Properties: The Bessel functions can be related to each other through recursive relations or differentiations.

Recursive Relations:

$$x J_{p+1}(x) = 2p J_p(x) - x J_{p-1}(x)$$

Differentiation Relations:

$$\frac{d}{dx}\left[x^{-p} J_p(x)\right] = -x^{-p} J_{p+1}(x)$$

$$\frac{d}{dx}\left[x^p J_p(x)\right] = x^p J_{p-1}(x)$$

$$\frac{d}{dx}\left[J_p(x)\right] = \frac{1}{2}\left[J_{p-1}(x) - J_{p+1}(x)\right].$$

Like with the Legendre polynomials, the Bessel functions $\{J_p(x)\}$ correspond to the coefficients of a power series expansion of the *generating function*

$$g(x,t) = \sum_{p=-\infty}^{\infty} J_p(x) t^p$$

which has a closed form given by

$$g(x,t) = e^{x(t-1/t)/2}.$$

To obtain the expressions for the Bessel functions, we use the series expansion of the exponential

$$e^{xt/2} e^{-x/(2t)} = \sum_{n=0} \frac{1}{n!}\left(\frac{xt}{2}\right)^n \sum_{m=0} \frac{(-1)^m}{m!}\left(\frac{x}{2t}\right)^n$$

$$= \sum_n \sum_m \frac{(-1)^m}{n! m!}\left(\frac{x}{2}\right)^{n+m} t^{n-m}$$

By substituting one of the summation indices with $p = n - m$,

$$e^{xt/2} e^{-x/(2t)} = \sum_{p=-\infty}^{\infty}\left(\sum_{m=\max(0,-p)} \frac{(-1)^m}{(p+m)! m!}\left(\frac{x}{2}\right)^{p+2m}\right) t^p.$$

Fourier Series

1 Lecture 19: Fourier Series for 2π-Periodicity

Periodic functions are used to represent phenomena that repeat themselves periodically. Many natural processes, like oscillations, vibrations, and waves, follow repeating patterns. We find many examples of this, from biological clocks—such as the firing of neurons or the beating of the heart—to the precise ticking of atomic clocks. These cyclic patterns are all governed by periodic behavior, which allows us to predict and analyze their behavior over time.

In this lecture, we focus on a periodic function $f(x)$ of a single variable x. A periodic function has the property that it returns the same value when x is shifted by a whole period. When the period is of length 2π, this means that

$$f(x) = f(x + 2k\pi), \text{ for every } x \text{ and } k = 0, \pm 1, \pm 2, \ldots$$

Any periodic function can be decomposed into a superposition of the fundamental periodic functions which are the trigonometric functions $\sin(kx)$ and $\cos(kx)$.

Definition 1 **Fourier series** is an infinite series representation of periodic functions in terms of trigonometric functions

$$\boxed{f(x) = \frac{a_0}{2} + \sum_{n=1}^{\infty} a_n \cos\left(n\frac{\pi x}{L}\right) + \sum_{n=1}^{\infty} b_n \sin\left(n\frac{\pi x}{L}\right)}$$

where $f(x)$ is defined on the **basic interval** of length $2L$ and repeats itself indefinitely. The coefficients a_n and b_n are determined by $f(x)$ through the orthogonality condition of the Fourier modes.

The original version of the chapter has been revised. A correction to this chapter can be found at
https://doi.org/10.1007/978-3-031-77053-1_7

1.1 Functions with 2π Periodicity

1.1.1 Basic Interval $[-\pi, \pi]$

To begin with, we consider functions with the natural 2π periodicity of the trigonometric functions

$$\sin(x) = \sin(x + 2\pi k), \quad \cos(x) = \cos(x + 2\pi k), \quad k \in \mathbb{Z}.$$

We also take the <u>basic interval</u> over which the function repeats itself to be $x \in [-\pi, \pi]$. The function $f(x)$ may be quite arbitrary within the basic interval, and may have cups and discontinuities. In this respect, Fourier series can handle more "difficult" functions than power (Taylor) series which can only represent differentiable functions.

Within this basic interval, i.e. $x \in [-\pi, \pi]$, the Fourier modes $\sin(x)$ and $\cos(x)$ represent the *fundamental modes* or the first harmonic, and $\sin(nx)$ and $\cos(nx)$ ($n \geq 2$) are the other *harmonics* which capture the rapidly-varying regions in $f(x)$.

Definition 2 (*Fourier coefficients*) The Fourier series expansion of a function with 2π periodicity is

$$\boxed{f(x) = \frac{a_0}{2} + \sum_{n=1}^{\infty} a_n \cos(nx) + \sum_{n=1}^{\infty} b_n \sin(nx),}$$

where the *Fourier coefficients* a_n and b_n are determined by $f(x)$ using the orthogonality condition of the Fourier modes on the basic interval $[-\pi, \pi]$.

Orthogonality of Trigonometric Functions: The sequence of Fourier modes,

$$\{\cos(nx), \sin(nx)\}_{n=0}^{\infty}$$

is a set of orthogonal functions on the basic interval $[-\pi, \pi]$. This implies the following orthogonality conditions. One such condition is that

$$\boxed{\int_{-\pi}^{\pi} dx \, \sin(nx) \cos(mx) = 0.}$$

We can see this quickly since we have an odd integrand over a symmetric interval

$$2\sin(nx)\cos(mx) = \sin((n+m)x) + \sin((n-m)x).$$

On the other hand, the integral

1 Lecture 19: Fourier Series for 2π-Periodicity

$$\boxed{\int_{-\pi}^{\pi} dx \sin(nx) \sin(mx) = \pi \delta_{m,n}}.$$

We see this quickly using that

$$2 \sin(nx) \sin(mx) = \cos((n-m)x) - \cos((n+m)x).$$

Similarly,

$$\boxed{\int_{-\pi}^{\pi} dx \cos(nx) \cos(mx) = \pi \delta_{m,n}}.$$

By taking half of the average of $f(x)$ in the basic interval $[-\pi, \pi]$, we find the coefficient a_0 as follows

$$a_0 = \frac{1}{\pi} \int_{-\pi}^{\pi} dx f(x)$$

$$= \frac{1}{\pi} \int_{-\pi}^{\pi} dx \frac{a_0}{2} + \frac{1}{\pi} \sum_{n=1}^{\infty} \int_{-\pi}^{\pi} dx a_n \cos(nx) + \frac{1}{\pi} \sum_{n=1}^{\infty} \int_{-\pi}^{\pi} dx b_n \sin(nx)$$

where the other terms in the series vanish. In general, by integrating $f(x) \cos(kx)$ over the basic interval, we extract the a_k coefficient of the Fourier series as follows:

$$\frac{1}{\pi} \int_{-\pi}^{\pi} dx f(x) \cos(kx)$$

$$= \frac{a_0}{2\pi} \int_{-\pi}^{\pi} \cos(kx) dx + \sum_{n=1}^{\infty} \frac{a_n}{\pi} \int_{-\pi}^{\pi} dx \cos(nx) \cos(kx)$$

$$+ \sum_{n=1}^{\infty} \frac{b_n}{\pi} \int_{-\pi}^{\pi} dx \sin(nx) \cos(kx)$$

$$= \sum_{n=1}^{\infty} a_n \delta_{n,k}$$

$$= a_k.$$

Similarly, the integral of $f(x) \sin(kx)$ over the basic interval gives the b_k coefficient of the Fourier series:

$$b_k = \frac{1}{\pi} \int_{-\pi}^{\pi} dx f(x) \sin(kx)$$

$$= \frac{a_0}{2\pi} \int_{-\pi}^{\pi} \sin(kx)dx + \sum_{n=1}^{\infty} \frac{a_n}{\pi} \int_{-\pi}^{\pi} dx \cos(nx)\sin(kx)$$

$$+ \sum_{n=1}^{\infty} \frac{b_n}{\pi} \int_{-\pi}^{\pi} dx \sin(nx)\sin(kx)$$

$$= \sum_{n=1}^{\infty} b_n \delta_{n,k}$$

$$= b_k$$

To summarize, the Fourier coefficients corresponding to a function defined on the basic interval $[-\pi, \pi]$ are given by

$$\boxed{a_k = \frac{1}{\pi} \int_{-\pi}^{\pi} dx f(x) \cos(kx), \quad b_k = \frac{1}{\pi} \int_{-\pi}^{\pi} dx f(x) \sin(kx).}$$

Example 1 To illustrate the point that Fourier series can handle functions with discontinuities, let us consider the step (Heaviside) function defined on the basic interval $[-\pi, \pi]$ with the jump at $x = 0$, hence

$$f(x) = \begin{cases} 1, & -\pi < x < 0 \\ 0, & 0 < x < \pi \end{cases} \tag{1}$$

and $f(x)$ is then extended over the real axis with 2π-periodicity as illustrated in Fig. 1. To find its Fourier series, we need to compute the coefficients. Thus,

Fig. 1 Periodic function given by Eq. 1

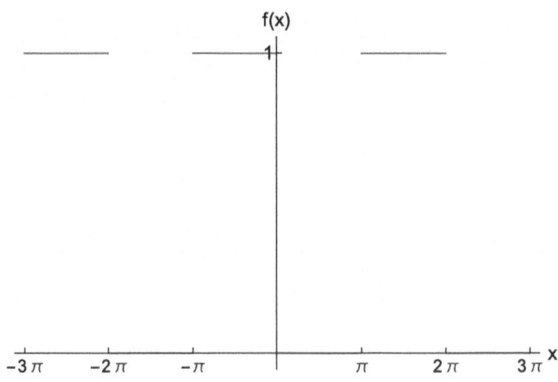

1 Lecture 19: Fourier Series for 2π-Periodicity

$$a_0 = \frac{1}{\pi} \int_{-\pi}^{\pi} dx f(x)$$

$$= \frac{1}{\pi} \int_{-\pi}^{0} dx$$

$$= 1.$$

The other coefficients are

$$a_k = \frac{1}{\pi} \int_{-\pi}^{\pi} dx f(x) \cos(kx)$$

$$= \frac{1}{\pi} \int_{-\pi}^{0} \cos(kx) dx$$

$$= \frac{1}{k\pi} \sin(kx) \Big|_{-\pi}^{0} = 0, k > 0,$$

and

$$b_k = \frac{1}{\pi} \int_{-\pi}^{\pi} dx f(x) \sin(kx)$$

$$= \frac{1}{\pi} \int_{-\pi}^{0} \sin(kx) dx = -\frac{1}{k\pi} \cos(kx) \Big|_{-\pi}^{0}$$

$$= -\frac{1}{k\pi} [1 - (-1^k)].$$

Thus, the Fourier series is given by

$$f(x) = \frac{1}{2} - \frac{2}{\pi} \sin(x) - \frac{2}{3\pi} \sin(3x) - \frac{2}{5\pi} \sin(5x) - \cdots$$

$$= \frac{1}{2} - \frac{1}{\pi} \sum_{k=1}^{\infty} \frac{1}{k} [1 - (-1)^k] \sin(kx).$$

Notice that all the odd terms in the series vanishes. Hence, we can simplify the expression for $k = 2n + 1$ as

$$f(x) = \frac{1}{2} - \frac{1}{\pi} \sum_{n=0}^{\infty} \frac{2}{2n+1} \sin[(2n+1)x].$$

Figure 2 shows the finite sum of this Fourier expansion for $N = 1, N = 2, N = 13, N = 100$.

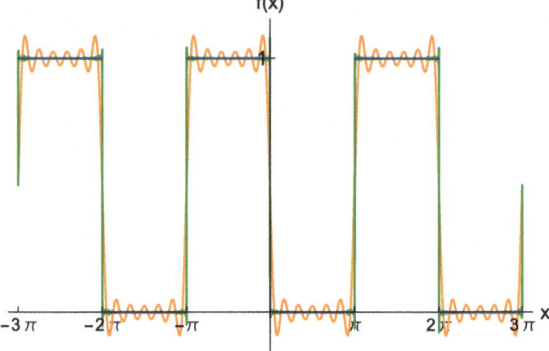

Fig. 2 Truncated Fourier series from Eq. 1 for $N = 5$, $N = 400$. Notice that the Fourier series overshoots around discontinuities

Gibbs Phenomenon: When we use a Fourier series to represent functions with discontinuities, an interesting phenomenon occurs. As we add more terms to the series, it converges to the function in smooth regions. At the discontinuity, the Fourier series goes through the midpoint of the jump and overshoots on both sides. This happens because the Fourier series takes longer to converge at these discontinuities, causing a net mismatch, no matter how many terms are included in the series.

Definition 3 Dirichlet conditions provide us with *sufficient* conditions on $f(x)$ such that the Fourier series converges to $f(x)$. These conditions are:

1. $f(x)$ has finite number of extreme points (min, max) in the basic interval, e.g. $[-\pi, \pi]$
2. $f(x)$ has finite number of discontinuities (finite jumps) in the basic interval
3. $f(x)$ is bounded, $\int_{-\pi}^{\pi} |f(x)| dx < 2\pi M$, where M is the maximum value of $f(x)$ in the basic interval.

At a discontinuity point x_0 in $f(x)$, the Fourier series converges to the midpoint

$$\lim_{\epsilon \to 0} \frac{1}{2} [f(x_0 + \epsilon) + f(x_0 - \epsilon)]$$

There are exceptions to this rule, i.e. there may be functions that do not fulfill the Dirichlet conditions and still have a convergent Fourier expansion.

Complex Form of the Fourier Series: Using the relations between the trigonometric functions and the complex exponential, we can express the Fourier series in the complex form. For the basic $[-\pi, \pi]$, this is given by

$$\boxed{f(x) = \sum_{n=-\infty}^{\infty} c_n e^{inx}.} \quad (2)$$

1 Lecture 19: Fourier Series for 2π-Periodicity

The **Fourier coefficients** are determined as

$$\boxed{c_k = \frac{1}{2\pi}\int_{-\pi}^{\pi} dx f(x) e^{-ikx},}$$

using the orthogonality relations of the complex exponential modes

$$\int_{-\pi}^{\pi} dx e^{inx} e^{-imx} = 2\pi \delta_{n,m}.$$

Multiplying on both sides of Eq. (2) with e^{-ikx} and integrating over the basic interval, we pick out the c_k coefficient of the series as

$$\frac{1}{2\pi}\int_{-\pi}^{\pi} dx f(x) e^{-ikx} = \frac{1}{2\pi}\sum_{n=-\infty}^{\infty} c_n \int_{-\pi}^{\pi} e^{inx} e^{-ikx}$$

$$= \sum_n c_n \delta_{n,k}$$

$$= c_k.$$

Example 2 For the function in Eq. (1), the zeroth order coefficient is

$$c_0 = \frac{1}{2\pi}\int_{-\pi}^{\pi} dx f(x) = \frac{1}{2\pi}\int_{-\pi}^{0} dx = \frac{1}{2}$$

and the higher order ones are

$$c_k = \frac{1}{2\pi}\int_{-\pi}^{\pi} dx f(x) e^{-ikx}$$

$$= -\frac{1}{2ik\pi} e^{-ikx}\Big|_{-\pi}^{0}$$

$$= -\frac{1}{2ik\pi}[1 - (e^{-i\pi})^k]$$

$$= -\frac{1}{2ik\pi}[1 - (-1)^k].$$

Thus, the Fourier series in the complex exponential form reads as

$$f(x) = \frac{1}{2} + \frac{i}{\pi}\sum_{k=-\infty}^{+\infty} \frac{1}{2k}[1 - (-1)^k] e^{ikx}.$$

To reduce it to the trigonometric form, we use the Euler's formula for the complex exponential

$$f(x) = \frac{1}{2} + \frac{i}{\pi} \sum_{k=-\infty}^{+\infty} \frac{1}{2k}[1 - (-1)^k]\cos(kx) - \frac{1}{\pi} \sum_{k=-\infty}^{+\infty} \frac{1}{2k}[1 - (-1)^k]\sin(kx).$$

Notice that the $\cos(kx)$ terms vanish because each term with positive k is cancelled out by a corresponding term with negative k. On the other hand, $\sin(kx)/k = \sin(-kx)/(-k)$ is even with respect to k, thus the sine series reduces to twice the series over positive k'. Hence,

$$f(x) = \frac{1}{2} - \frac{1}{\pi} \sum_{k=1}^{+\infty} \frac{1}{k}[1 - (-1)^k]\sin(kx).$$

Notice that the Fourier coefficients have a slow decay as $c_k \sim 1/k$, because the function has points of discontinuity. Thus, the derivative of $f(x)$ cannot be represented by a convergent Fourier series.

1.1.2 Other Basic Intervals of 2π Length

We can use the same Fourier modes $\{\cos(nx), \sin(nx)\}$ or $\{e^{inx}\}$ to expand 2π-periodic functions defined on shifted basic intervals, e.g. $[0, 2\pi]$, $[3\pi, 5\pi]$, etc. While the expansion is formally the same, the coefficients $\{a_n, b_n\}$ or, equivalently, $\{c_n\}$ are unique to the basic interval, and they represent different periodic functions! For instance, let us take the function

$$f(x) = x^3$$

and extend it periodically with the basic interval $[-\pi, \pi]$ as illustrated in Fig. 3. Now, let us extend the same function on the basic internal $[0, 2\pi]$. This generates a completely different periodic signal as shown in Fig. 3.

The important difference is that the orthogonality condition of the Fourier modes applies to different integration intervals. For the basic interval $[0, 2\pi]$, the orthogonality condition is

$$\int_0^{2\pi} dx\, e^{inx} e^{-imx} = 2\pi \delta_{n,m}$$

This implies that when $f(x)$ is defined over a basic interval $[0, 2\pi]$, the Fourier coefficients c_n are given by

$$\boxed{c_n = \frac{1}{2\pi} \int_0^{2\pi} dx\, f(x) e^{-inx}}$$

1 Lecture 19: Fourier Series for 2π-Periodicity

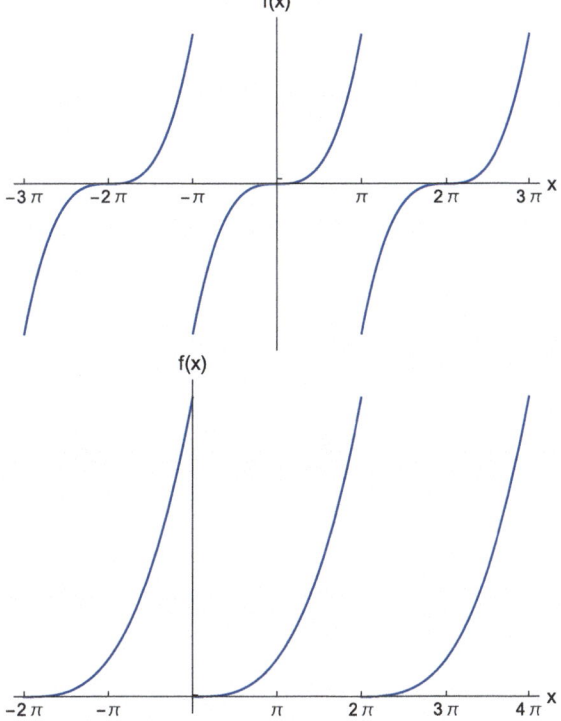

Fig. 3 Two different periodic functions starting from $f(x) = x^3$ in different basic intervals: (left) $x \in [-\pi, \pi]$, (right) $x \in [0, 2\pi]$

and, equivalently, the Fourier coefficients of the trigonometric form are

$$a_n = \frac{1}{\pi} \int_0^{2\pi} dx f(x) \cos(nx), \quad b_n = \frac{1}{\pi} \int_0^{2\pi} dx f(x) \sin(nx)$$

Example 3 Let us take the $f(x) = x^3$ and derive the corresponding Fourier series

$$f(x) = \sum_{n=-\infty}^{\infty} c_n e^{inx}$$

for the basic intervals $[-\pi, \pi]$ and $[0, 2\pi]$.

Basic Interval $[-\pi, \pi]$: The zeroth order coefficient is

$$c_0 = \frac{1}{2\pi} \int_{-\pi}^{\pi} dx x^3 = 0$$

For $n \neq 0$, the integral is evaluated by successive integrations by parts and given by

$$c_n = \frac{1}{2\pi} \int_{-\pi}^{\pi} dx\, x^3 e^{-inx} = i\frac{(-1)^n}{n^3}(\pi^2 n^2 - 6).$$

Hence, the Fourier series reads as

$$f(x) = i \sum_{n=\pm 1}^{\infty} \frac{(-1)^n}{n^3}(\pi^2 n^2 - 6) e^{inx}$$

$$= i \sum_{n=1}^{\infty} \frac{(-1)^n}{n^3}(\pi^2 n^2 - 6)\left(e^{inx} - e^{-inx}\right)$$

$$= 2 \sum_{n=1}^{\infty} \frac{(-1)^{n+1}}{n^3}(\pi^2 n^2 - 6) \sin(nx).$$

Basic Interval $[0, 2\pi]$: The zeroth order coefficient is

$$c_0 = \frac{1}{2\pi} \int_0^{2\pi} dx\, x^3 = 2\pi^3$$

while the other coefficients for $n \neq 0$

$$c_n = \frac{1}{2\pi} \int_0^{2\pi} dx\, x^3 e^{-inx}$$

$$= \frac{2}{n^3}(2i\pi^2 n^2 + 3\pi n - 3i) = 2i\left(\frac{2\pi^2}{n} - \frac{3}{n^3}\right) + \frac{6\pi}{n^2}.$$

Hence, the Fourier series reads as

$$f_{series}(x) = 2\pi^3 + 2i \sum_{n=\pm 1}^{\infty} \left(\frac{2\pi^2}{n} - \frac{3}{n^3}\right) e^{inx} + 6\pi \sum_{n=\pm 1}^{\infty} \frac{1}{n^2} e^{inx}$$

$$= 2\pi^3 + 2i \sum_{n=1}^{\infty} \left(\frac{2\pi^2}{n} - \frac{3}{n^3}\right)\left(e^{inx} - e^{-inx}\right) + 6\pi \sum_{n=1}^{\infty} \frac{1}{n^2}\left(e^{inx} + e^{-inx}\right)$$

$$= 2\pi^3 - 4 \sum_{n=1}^{\infty} \left(\frac{2\pi^2}{n} - \frac{3}{n^3}\right) \sin(nx) + 12\pi \sum_{n=1}^{\infty} \frac{1}{n^2} \cos(nx).$$

Something interesting happens at $x = 0$ where the periodic function has a jump of size $(2\pi)^3$. The Fourier series $f_{series}(x = 0)$ evaluated at this point is given by

$$f_{series}(x=0) = 2\pi^3 + 12\pi \sum_{n=1}^{\infty} \frac{1}{n^2},$$

and because $x = 0$ is a discontinuity point, it also interpolates to the midpoint value of the jump, hence

$$f_{series}(x=0) = \frac{(2\pi)^3}{2} = 4\pi^3.$$

Combining these two expressions at $x = 0$, we find a closed expression for the infinite sum

$$\sum_{n=1}^{\infty} \frac{1}{n^2} = \frac{\pi^2}{6}.$$

This technique is useful for finding closed expressions for other infinite sums.

2 Lecture 20: Fourier Series for 2L-Periodicity

For $f(x)$ with a generic $2L$ periodicity, it follows that

$$f(x) = f(x + 2kL), \quad k \in \mathcal{Z},$$

and the corresponding Fourier modes need to have the same periodicity. This is achieved by re-scaling x as

$$x \to \frac{\pi x}{L},$$

such that the trigonometric Fourier modes

$$\{\cos\left(n\frac{\pi x}{L}\right), \sin\left(n\frac{\pi x}{L}\right)\}$$

have a $2L$ period. That is

$$\cos\left(n\frac{\pi(x+2L)}{L}\right) = \cos\left(n\frac{\pi x}{L} + 2n\pi\right) = \cos\left(n\frac{\pi x}{L}\right) \tag{3}$$

$$\sin\left(n\frac{\pi(x+2L)}{L}\right) = \sin\left(n\frac{\pi x}{L} + 2n\pi\right) = \sin\left(n\frac{\pi x}{L}\right) \tag{4}$$

These Fourier modes form a complete basis for the Fourier expansion, given by

$$\boxed{f(x) = \frac{a_0}{2} + \sum_{n=1}^{\infty} a_n \cos\left(n\frac{\pi x}{L}\right) + \sum_{n=1}^{\infty} b_n \sin\left(n\frac{\pi x}{L}\right).}$$

Similarly, the complex exponentials with $2L$-periodicity are $\{e^{in\frac{\pi x}{L}}\}$, where for each n

$$e^{i\frac{n\pi}{L}(x+2L)} = e^{i\frac{n\pi}{L}x}e^{i2n\pi} = e^{i\frac{n\pi}{L}x}.$$

The complex exponential form of the Fourier series reads as

$$\boxed{f(x) = \sum_{n=-\infty}^{\infty} c_n e^{in\pi x/L}}$$

The Fourier coefficients are determined by the orthogonality relations of the Fourier modes in the basic interval of length $2L$.

Basic Interval $[-L, L]$: The orthogonality condition of the complex Fourier modes over the basic interval $[-\pi, \pi]$ is

$$\boxed{\frac{1}{2L}\int_{-L}^{L} dx\, e^{in\pi x/L} e^{-im\pi x/L} = \delta_{n,m}}$$

For $n = m$, the integral is

$$\frac{1}{2L}\int_{-L}^{L} dx = 1,$$

while for any $n \neq m$,

$$\frac{1}{2L}\int_{-L}^{L} e^{i(n-m)\pi x/L} dx = \frac{1}{2\pi i(n-m)}\left[e^{i(n-m)\pi} - e^{-i(n-m)\pi}\right] = 0.$$

The Fourier coefficients c_n follow by integrating the function multiplied the appropriate Fourier mode over the basic interval,

$$\frac{1}{2L}\int_{-L}^{L} dx\, f(x) e^{-im\pi x/L} = \sum_n c_n \frac{1}{2L}\int_{-L}^{L} dx\, e^{in\pi x/L} e^{-im\pi x/L} \quad (5)$$

$$= \sum_n c_n \delta_{n,m} \quad (6)$$

$$= c_m \quad (7)$$

2 Lecture 20: Fourier Series for 2L-Periodicity

Thus,

$$c_n = \frac{1}{2L} \int_{-L}^{L} dx f(x) e^{-in\pi x/L}$$

Similarly, the Fourier coefficients for the trigonometric form follow as

$$a_n = \frac{1}{L} \int_{-L}^{L} dx f(x) \cos\left(n\frac{\pi x}{L}\right), \quad b_n = \frac{1}{L} \int_{-L}^{L} dx f(x) \sin\left(n\frac{\pi x}{L}\right)$$

Example 4 Let us as consider this function

$$f(x) = \begin{cases} 1, & -1 < x < 0 \\ 0, & 0 < x < 1 \end{cases}$$

defined in the basic interval $[-1, 1]$ and extended over the real axis. It has the periodicity of $2L = 2$, hence the Fourier modes $\{e^{in\pi x}\}$ have the same period. The complex Fourier expansion reads as

$$f(x) = \sum_{n=-\infty}^{\infty} c_n e^{in\pi x}$$

with the Fourier coefficients computed as

$$c_n = \frac{1}{2} \int_{-1}^{1} dx f(x) e^{-in\pi x}$$

$$= \frac{1}{2} \int_{-1}^{0} dx e^{-in\pi x}$$

$$= \begin{cases} \frac{1}{2}, & n = 0 \\ \frac{i}{2n\pi}[1 - (-1)^n], & n \neq 0 \end{cases}$$

Thus, the Fourier expansion reduces to

$$f(x) = \frac{1}{2} + \frac{i}{\pi} \sum_{n=\pm\text{odd}} \frac{1}{n} e^{in\pi x}$$

$$= \frac{1}{2} - \frac{2}{\pi} \sum_{n=\text{odd}} \frac{\sin(n\pi x)}{n} \quad (8)$$

Figure 4 shows Fourier series truncated after the first few terms.

Fig. 4 Truncated Fourier series from Eq. 8

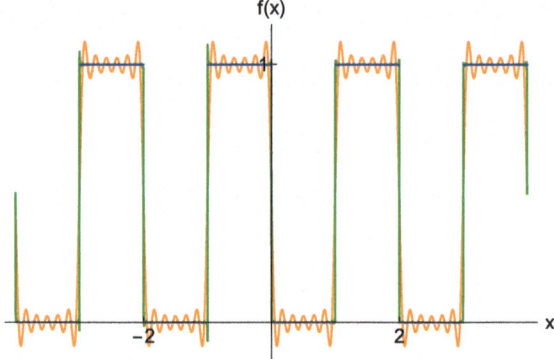

Basic Interval $[0, 2L]$: The corresponding orthogonality conditions for this basic interval are

$$\boxed{\frac{1}{2L} \int_0^{2L} dx\, e^{in\pi x/L} e^{-im\pi x/L} = \delta_{n,m}}$$

Example 5 Let us consider this function

$$f(x) = x, \quad x \in [0, 1]$$

The corresponding Fourier modes with the same period $2L = 1$ are $\{e^{i2n\pi x}\}$ since (Fig. 5)

$$e^{i2n\pi(x+1)} = e^{i2n\pi x} e^{i2n\pi} = e^{i2n\pi x}.$$

Therefore, the complex Fourier expansion reads as

Fig. 5 Truncated Fourier series from Eq. (12)

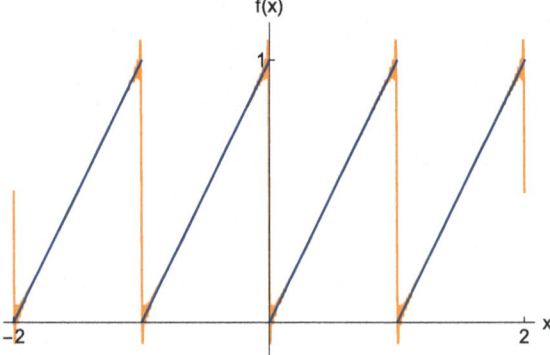

$$f(x) = \sum_{n=-\infty}^{\infty} c_n e^{2\pi i n x},$$

where the coefficients are computed from the orthogonality condition of the Fourier modes on the [0, 1] basic interval as

$$c_n = \int_0^1 dx\, x e^{-2\pi i n x}$$

For $n = 0$:

$$c_0 = \int_0^1 dx\, x = \frac{1}{2}$$

For $n \neq 0$:

$$c_n = \frac{i}{2n\pi} \int_0^1 dx\, x \frac{d}{dx} e^{-i2n\pi x} \tag{9}$$

$$= \frac{i}{2n\pi} \left[x e^{-i2n\pi x} \right]_0^1 - \frac{i}{2n\pi} \int_0^1 dx\, e^{-i2n\pi x} \tag{10}$$

$$= \frac{i}{2n\pi} e^{-i2n\pi} = \frac{i}{2n\pi}. \tag{11}$$

Thus, the Fourier series becomes

$$f(x) = \frac{1}{2} + \frac{i}{2\pi} \sum_{n\neq 0} \frac{1}{n} e^{i2n\pi x}$$

$$= \frac{1}{2} - \frac{1}{\pi} \sum_{n=1}^{\infty} \frac{\sin(2\pi n x)}{n}. \tag{12}$$

2.1 Even and Odd Extensions

For a function $f(x)$ defined over a basic interval $[0, L]$, we may choose how to extend it periodically across the entire real axis. An important advantage of this flexibility is that we can introduce even or odd symmetries when extending the function beyond the basic interval.

Fig. 6 Fourier series of the even extension from Eq. (16)

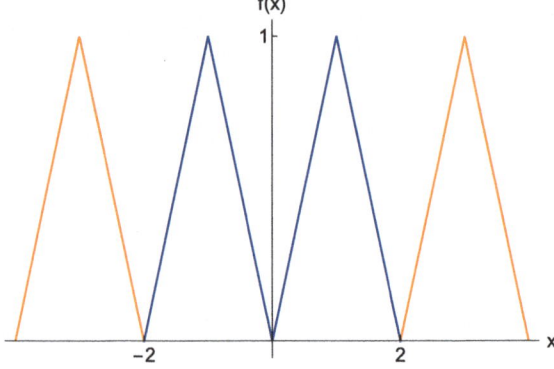

Even Extension with a Period of $2L$: We may construct a periodic function on the basic interval $[-L, L]$ as

$$\tilde{f}_{even}(x) = \begin{cases} f(-x), & -L < x < 0 \\ f(x), & 0 < x < L \end{cases} \quad (13)$$

This symmetry must also hold for the Fourier expansion, which means that only the even Fourier modes have non-zero coefficients $b_n = 0$ because

$$\int_{-L}^{L} dx\, \tilde{f}_{even}(x) \sin\left(n\frac{\pi x}{L}\right) = 0.$$

Thus, an even function has a **cosine series**:

$$\tilde{f}_{even}(x) = \frac{a_0}{2} + \sum_{n=1}^{\infty} a_n \cos\left(n\frac{\pi x}{L}\right),$$

where the Fourier coefficients are given by

$$a_n = \frac{1}{L} \int_{-L}^{L} dx\, \tilde{f}_{even}(x) \cos\left(n\frac{\pi x}{L}\right) \quad (14)$$

$$= \frac{2}{L} \int_{0}^{L} dx\, f(x) \cos\left(n\frac{\pi x}{L}\right). \quad (15)$$

Example 6 Let us consider the previous function $f(x) = x$ on the basic interval $[0, 1]$. We have computed its Fourier transform when the function is replicated over

2 Lecture 20: Fourier Series for 2L-Periodicity

the real axis with periodicity $2L = 1$. Now, let us construct its *even* extension which is a periodic function over the basic interval $[-1, 1]$ (Fig. 6):

$$\tilde{f}_{even}(x) = \begin{cases} -x, & -L < x < 0 \\ x, & 0 < x < L \end{cases}$$

The coefficients of the *cosine series*, with

$$a_0 = 2 \int_0^1 dx\, x = 1$$

and

$$a_n = 2 \int_0^1 dx\, x \cos(n\pi x) = -\frac{2}{n^2\pi^2}[1 - (-1)^n]$$

such that resulting cosine series is

$$\tilde{f}_{even}(x) = \frac{1}{2} - \frac{4}{\pi^2} \sum_{n=odd}^{\infty} \frac{1}{n^2} \cos(n\pi x). \tag{16}$$

Odd Extension with a Period $2L$: Similarly, given $f(x)$ on a basic interval $[0, L]$, we may construct its odd extension in the basic interval $[-L, L]$ as

$$\tilde{f}_{odd}(x) = \begin{cases} -f(-x), & -L < x < 0 \\ f(x), & 0 < x < L \end{cases}$$

which has the *sine series expansion*

$$\tilde{f}_{odd}(x) = \sum_{n=1}^{\infty} b_n \sin(n\pi x),$$

with the Fourier coefficients

$$b_n = \frac{1}{L} \int_{-L}^{L} dx\, \tilde{f}_{odd}(x) \sin\left(n\frac{\pi x}{L}\right)$$

$$= \frac{2}{L} \int_0^L dx\, f(x) \sin\left(n\frac{\pi x}{L}\right).$$

Note that the Fourier coefficients $a_n = 0$ because

$$\int_{-L}^{L} dx\, \tilde{f}_{odd}(x) \cos\left(n\frac{\pi x}{L}\right) = 0$$

due to the odd symmetry of the integrand (Fig. 7).

Example 7 Let us take the same function $f(x) = x$ with the basic interval $[0, 1]$. We now construct its *odd* extension over the basic interval $[-1, 1]$:

$$\tilde{f}(x) = \begin{cases} x, & -L < x < 0 \\ x, & 0 < x < L \end{cases}$$

The odd Fourier modes with $2L$-periodicity are $\{\sin(n\pi x)\}$, and their corresponding coefficients are computed as

$$b_n = 2\int_0^1 dx\, x \sin(n\pi x) = -\frac{2}{n\pi}(-1)^n.$$

Hence, the sine series reads as

$$\tilde{f}(x) = -\frac{2}{\pi} \sum_{n=1}^{\infty} \frac{(-1)^n}{n} \sin(n\pi x). \tag{17}$$

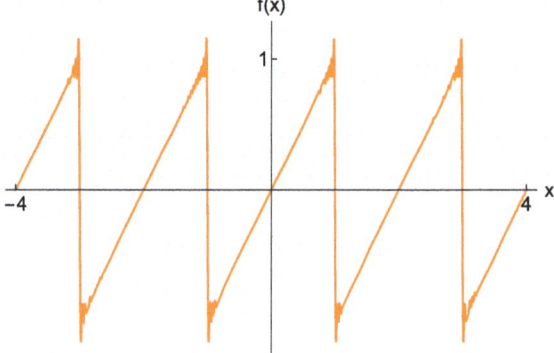

Fig. 7 Truncated sin series from Eq. (17)

2.2 Completeness Relation

Another important property of the Fourier series is the relation satisfied by the coefficients of the Fourier modes.

Theorem 1 (Parseval's theorem) *The set of Fourier modes $\{\cos(n\pi x/L), \sin(n\pi x/L)\}$ or equivalently $\{e^{in\pi x/L}\}$ is a complete basis for the Fourier expansion*

$$f(x) = \frac{a_0}{2} + \sum_{n=1}^{\infty} a_n \cos\left(n\frac{\pi x}{L}\right) + \sum_{n=1}^{\infty} b_n \cos\left(n\frac{\pi x}{L}\right)$$

or, equivalently,

$$f(x) = \sum_{n=-\infty}^{\infty} c_n e^{in\pi x/L},$$

when the average of $|f(x)|^2$ over the basic interval $[-L, L]$ is determined by the ***completeness relation***

$$\frac{1}{2L}\int_{-L}^{L} dx |f(x)|^2 = \left(\frac{a_0}{2}\right)^2 + \frac{1}{2}\sum_{n=1}^{\infty} |a_n|^2 + \frac{1}{2}\sum_{n=1}^{\infty} |b_n|^2$$

or, equivalently,

$$\frac{1}{2L}\int_{-L}^{L} dx |f(x)|^2 = \sum_{n=-\infty}^{\infty} |c_n|^2$$

To show this, we use the orthogonality relation. For the complex representation of the Fourier series, this follows as

$$\frac{1}{2L}\int_{-L}^{L} dx |f(x)|^2 = \frac{1}{2L}\int_{-L}^{L} dx \sum_n \sum_m c_n c_m^* e^{in\pi x/L} e^{-im\pi x/L}$$

$$= \sum_n \sum_m c_n c_m^* \frac{1}{2L}\int_{-L}^{L} dx\, e^{i(n-m)\pi x/L}$$

$$= \sum_n \sum_m c_n c_m^* \delta_{n,m}$$

$$= \sum_n |c_n|^2$$

Similarly, for the trigonometric representation of the Fourier series, the only non-vanishing terms are

$$\frac{1}{2L}\int_{-L}^{L} dx |f(x)|^2 = \left(\frac{a_0}{2}\right)^2 + \sum_{n=1}^{\infty}\sum_{m=1}^{\infty} a_n a_m \frac{1}{2L}\int_{-L}^{L} dx \cos\left(\frac{n\pi x}{L}\right)\cos\left(\frac{m\pi x}{L}\right)$$

$$+ \sum_{n=1}^{\infty}\sum_{m=1}^{\infty} b_n b_m \frac{1}{2L}\int_{-L}^{L} dx \sin\left(\frac{n\pi x}{L}\right)\sin\left(\frac{m\pi x}{L}\right)$$

$$= \left(\frac{a_0}{2}\right)^2 + \sum_{n=1}^{\infty} a_n^2 \frac{1}{2L}\int_{-L}^{L} dx \cos^2\left(\frac{n\pi x}{L}\right) + \sum_{n=1}^{\infty} b_n^2 \frac{1}{2L}\int_{-L}^{L} dx \sin^2\left(\frac{n\pi x}{L}\right)$$

$$= \left(\frac{a_0}{2}\right)^2 + \frac{1}{2}\sum_{n=1}^{\infty} a_n^2 + \frac{1}{2}\sum_{n=1}^{\infty} b_n^2.$$

For instance, the Fourier basis $\{\sin(2nx), \cos(2nx)\}$ is incomplete for basic interval of length 2π, because the fundamental mode ($n = 1$) has a wavelength that is π instead of 2π. Hence, the associated Fourier coefficients will not satisfy the completeness relation. The complete basis for the 2π-periodicity is $\{\sin(nx), \cos(nx)\}$.

Closed Form for Infinite Series: Parseval's theorem has a number of useful applications, one of them being that we may use it to find the sum of infinite series.

Example 8 Show that

$$\sum_{n=1}^{\infty} \frac{1}{n^2} = \frac{\pi^2}{6}$$

Solution: The terms in the infinite series correspond to the Fourier coefficients of the periodic function $f(x) = x$ on the basic interval $x \in [-1, 1]$. The corresponding Fourier modes that also have a period of 2 are $\{e^{in\pi x}\}$ and their corresponding coefficients follow as

$$c_n = \frac{1}{2}\int_{-1}^{1} dx\, x e^{-in\pi x}.$$

For $n = 0$, this implies

$$c_0 = \frac{1}{2}\int_{-1}^{1} dx\, x = 0,$$

and for $n \neq 0$, this is

2 Lecture 20: Fourier Series for 2L-Periodicity

$$c_n = \frac{i}{2n\pi} \int_{-1}^{1} dx \, x \frac{d}{dx} e^{-in\pi x}$$

$$= \frac{i}{2n\pi} \left[xe^{-in\pi x} \right]_{-1}^{1} - \frac{i}{2n\pi} \int_{-1}^{1} dx \, e^{-in\pi x}$$

$$= \frac{i}{2n\pi} \left[e^{in\pi} + e^{-in\pi} \right]$$

$$= \frac{i}{n\pi} \cos(n\pi) = (-1)^n \frac{i}{n\pi}.$$

Thus, the Fourier expansion is

$$f(x) = \sum_{n=-\infty}^{\infty} (-1)^n \frac{i}{n\pi} e^{in\pi x}.$$

We can then find a closed form of the infinite series through Parseval's identity, namely

$$\frac{1}{2} \int_{-1}^{1} dx |x|^2 = \frac{1}{\pi^2} \sum_{n \neq 0} \frac{1}{n^2}$$

$$= \frac{2}{\pi^2} \sum_{n=1}^{\infty} \frac{1}{n^2}.$$

The integral over the basic interval can be evaluated straightforwardly,

$$\frac{1}{2} \int_{0}^{1} dx \, x^2 = \frac{1}{6} |x|^3 \Big|_{-1}^{1} = \frac{1}{3},$$

and leads to the sum,

$$\sum_{n=1}^{\infty} \frac{1}{n^2} = \frac{\pi^2}{6}.$$

Integral Transforms

1 Lecture 21: Fourier Transform

An integral transform is a linear transformation of a function $f(x)$ from one function space to another through integration. This mapping is reversible, allowing us to recover the original function using an inverse integral transform. Two of the most commonly used integral transforms are the Fourier transform and the Laplace transform. These transforms are especially useful in solving differential equations.

In this lecture, we focus on the Fourier transform and its properties.

1.1 Fourier Transform

As a transformation, the Fourier transform takes a function $f(x)$ defined in the x-space and maps it into another function $\hat{f}(k)$ defined in the conjugate k-space.

$$g(k) \equiv \mathcal{F}[f(x)] \text{ is the Fourier transform of } f(x)$$

and

$$f(x) \equiv \mathcal{F}^{-1}[g(k)] \text{ is the inverse Fourier transform of } g(k).$$

They are a pair of conjugate functions.

The Fourier transform is an extension of the Fourier series to non-periodic functions. A function $f(x)$ that is periodic can be decomposed into **discrete** Fourier modes. A non-periodic function can also be represented by Fourier modes but with a **continuous** spectrum of wavenumbers. The Fourier modes $\{e^{ik_n x}\}$ corresponding to a periodic function have discrete wave-numbers $k_n = n\pi/L$ (where n is integer).

The original version of the chapter has been revised. A correction to this chapter can be found at https://doi.org/10.1007/978-3-031-77053-1_7

© The Author(s), under exclusive license to Springer Nature Switzerland AG 2025, corrected publication 2025 L. Angheluta, *Analytical Methods in Physics*, https://doi.org/10.1007/978-3-031-77053-1_5

However, when $f(x)$ is aperiodic, the basic interval extends over the whole real axis $L \to \infty$ such that the wavenumbers vary continuously.

Theorem 1 (Fourier transform theorem) *If $f(x)$ satisfies the Dirichlet's conditions on every finite interval, i.e.*

- *the function has finite number of cups or jumps*
- *its norm $\int_{-\infty}^{\infty} |f(x)|^2 dx$ is finite,*

then $f(x)$ has a well-defined Fourier transform given by

$$\boxed{f(x) = \int_{-\infty}^{\infty} dk g(k) e^{ikx}, \qquad g(k) = \frac{1}{2\pi} \int_{-\infty}^{\infty} dx f(x) e^{-ikx}} \qquad (1)$$

where

$$g(k) = \mathcal{F}[f(x)]$$

is called the **Fourier transform** of $f(x)$, and

$$f(x) = \mathcal{F}^{-1}[g(k)]$$

is the **inverse Fourier transform** of $g(k)$.

These form a pair of conjugate functions which can be transformed into one another. In equivalent formulations, there may be a different prefactor in front of the integrals for $f(x)$ and $g(x)$. *The important things is that the product of prefactors has to be 2π.* Equivalent transformations are

$$f(x) = \sqrt{\frac{1}{2\pi}} \int_{-\infty}^{\infty} dk g(k) e^{ikx}, \qquad g(k) = \sqrt{\frac{1}{2\pi}} \int_{-\infty}^{\infty} dx f(x) e^{-ikx}. \qquad (2)$$

The Dirichlet conditions are sufficient conditions and a good rule of thumb. However, there also functions which can be Fourier transformed even though they do not satisfy these conditions. The Dirac delta function is one such example.

To prove this, we use that $f(x)$ has a basic interval that extends over the whole real axis, namely that it has an infinite periodicity. For a set of discrete wavenumbers

$$k_n = n\frac{\pi}{L},$$

where n is an integer, the wavenumber increment is constant and given by

$$\Delta k \equiv k_{n+1} - k_n = \frac{\pi}{L}.$$

1 Lecture 21: Fourier Transform

In the limit of $L \to \infty$, the distance between successive wavenumbers becomes infinitesimally small, i.e. $\Delta k \to 0$. Let us use the definition of the coefficients c_n of the complex Fourier modes corresponding to a periodic function defined on the basic interval $[-L, L]$:

$$c_n = \frac{1}{2L} \int_{-L}^{L} du\, f(u) e^{-ik_n u} = \frac{\Delta k}{2\pi} \int_{-L}^{L} du\, f(u) e^{-ik_n u}.$$

Inserting this into the Fourier series of $f(x)$, we find that

$$f(x) = \sum_{n=-\infty}^{\infty} c_n e^{ik_n x}$$

$$= \sum_{n=-\infty}^{\infty} \frac{\Delta k}{2\pi} \left[\int_{-L}^{L} du\, f(u) e^{ik_n(x-u)} \right]$$

Thus, as $L \to \infty$, the sum has infinitely many terms and approaches an integral over the continuous variable k

$$f(x) = \frac{1}{2\pi} \int_{-\infty}^{\infty} dk \left[\int_{-\infty}^{\infty} du\, f(u) e^{ik(x-u)} \right]$$

$$= \frac{1}{2\pi} \int_{-\infty}^{\infty} dk \left[\int_{-\infty}^{\infty} du\, f(u) e^{-iku} \right] e^{ikx}$$

$$= \int_{-\infty}^{\infty} dk\, g(k) e^{ikx}$$

where

$$g(k) = \frac{1}{2\pi} \int_{-\infty}^{\infty} du\, f(u) e^{-iku}$$

$$= \frac{1}{2\pi} \int_{-\infty}^{\infty} dx\, f(x) e^{-ikx}.$$

Notice that in this derivation, we have the choice how we want to split the prefactor $\frac{1}{2\pi}$ between the two Fourier integrals. This leads to the two equivalent formulations. It is important to be aware of this and be consistent with your choice.

Example 1 Let us compute the Fourier transform of the Gaussian $f(x) = e^{-\alpha x^2}$.

Solution: The Gaussian is a smooth function and has the property that

$$\int_{-\infty}^{\infty} dx\, e^{-2\alpha x^2} = \sqrt{\frac{\pi}{2\alpha}}.$$

Its Fourier transform is given by

$$g(k) = \frac{1}{2\pi} \int_{-\infty}^{\infty} dx\, f(x) e^{-ikx}$$

$$= \frac{1}{2\pi} \int_{-\infty}^{\infty} dx\, e^{-\alpha x^2} e^{-ikx},$$

which can be integrated by completing the square in the exponent as

$$\alpha x^2 + ikx = \alpha x^2 + ikx + \frac{(ik)^2}{4\alpha} + \frac{k^2}{4\alpha} = \left(\sqrt{\alpha} x + \frac{ik}{2\sqrt{\alpha}}\right)^2 + \frac{k^2}{4\alpha}.$$

Inserting in the integral above, we have that

$$g(k) = \frac{1}{2\pi} e^{-k^2/(4\alpha)} \int_{-\infty}^{\infty} dx\, e^{-(\sqrt{\alpha} x + ik/(2\sqrt{\alpha}))^2}$$

$$= \frac{1}{2\pi} e^{-k^2/(4\alpha)} \sqrt{\frac{\pi}{\alpha}}$$

$$= \sqrt{\frac{1}{4\pi\alpha}} e^{-k^2/(4\alpha)}.$$

Thus, the Fourier transform of a Gaussian is also a Gaussian in the Fourier space.

Example 2 Let us now compute the Fourier transform of the exponential $f(x) = e^{-|x|}$.

Solution: The corresponding Fourier transform is given by

$$g(k) = \mathcal{F}\left(e^{-|x|}\right) = \frac{1}{2\pi} \int_{-\infty}^{\infty} dx\, f(x) e^{-ikx}$$

$$= \frac{1}{2\pi} \int_{-\infty}^{\infty} dx\, e^{-|x|} e^{-ikx}$$

1 Lecture 21: Fourier Transform

$$= \frac{1}{2\pi} \left[\int_{-\infty}^{0} dx e^{-(ik-1)x} + \int_{0}^{\infty} dx e^{-(ik+1)x} \right]$$

$$= \frac{1}{2\pi} \left[-\frac{1}{ik-1} + \frac{1}{ik+1} \right]$$

$$= \frac{1}{\pi} \frac{1}{1+k^2}$$

The inverse Fourier transform is

$$e^{-|x|} = \mathcal{F}^{-1}\left(\frac{1}{1+k^2}\right) = \frac{1}{\pi} \int_{-\infty}^{\infty} dk \frac{e^{ikx}}{1+k^2}$$

Fourier Transform of Dirac δ-Function: Let us know compute the Fourier transform of the Dirac delta $\delta(x-a)$.

Solution: Using the properties of the δ-function

$$\int_{-\infty}^{\infty} dx \delta(x-a) = 1$$

and

$$\int_{-\infty}^{\infty} dx \delta(x-a) f(x) = f(a),$$

we quickly find that the Fourier transform of $\delta(x-a)$ is

$$\mathcal{F}[\delta(x-a)] = \frac{1}{2\pi} \int_{-\infty}^{\infty} dx \delta(x-a) e^{-ikx} = \frac{1}{2\pi} e^{-ika}. \tag{3}$$

Thus,

$$\boxed{\delta(x-a) = \frac{1}{2\pi} \int_{-\infty}^{\infty} dk e^{ik(x-a)}.}$$

1.1.1 Fourier Transform of Derivatives

Let $f(x)$ and $\mathcal{F}[f(x)]$ be a pair of Fourier transforms. Then the Fourier transform of the derivative $f'(x)$ is determined by the Fourier transform of $f(x)$ by the following transformation

$$\boxed{\mathcal{F}[f'(x)] = ik\mathcal{F}[f(x)]}$$

In general, the Fourier transform of the n'th derivative of $f(x)$ is

$$\boxed{\mathcal{F}[f^{(n)}(x)] = (ik)^n \mathcal{F}[f(x)].}$$

To see this, let us compute the Fourier transform of $f'(x)$,

$$\mathcal{F}[f'(x)] = \frac{1}{2\pi} \int_{-\infty}^{\infty} dx f'(x) e^{-ikx}$$

$$= ik \frac{1}{2\pi} \int_{-\infty}^{\infty} dx f(x) e^{-ikx}$$

$$= ik\mathcal{F}[f(x)],$$

using the integration by parts and that $f(x)$ decays to zero at $\pm\infty$. For the second order derivative, we use again the integration by parts to move each d/dx onto e^{-ikx}:

$$\mathcal{F}[f''(x)] = ik \frac{1}{2\pi} \int_{-\infty}^{\infty} dx f'(x) e^{-ikx}$$

$$= ik\mathcal{F}[f'(x)]$$

$$= (ik)^2 \mathcal{F}[f(x)].$$

Thus, in general, the Fourier transform of the n'th derivative is determined by successive integration by parts,

$$\mathcal{F}[f^{(n)}(x)] = ik \frac{1}{2\pi} \int_{-\infty}^{\infty} dx f^{(n-1)}(x) e^{-ikx}$$

$$= ik\mathcal{F}[f^{(n-1)}(x)]$$

$$= (ik)^n \mathcal{F}[f(x)].$$

These properties are particularly useful in solving differential equations by transforming them into algebraic equations in the k-space.

1.1.2 Fourier Transform of Symmetric Functions

We may write the Fourier transforms in terms of trigonometric functions using the Euler's formula as

$$g(k) = \frac{1}{2\pi} \int_{-\infty}^{\infty} dx f(x) \cos(kx) - i \frac{1}{2\pi} \int_{-\infty}^{\infty} dx f(x) \sin(kx)$$

and the inverse transform as

$$f(x) = \int_{-\infty}^{\infty} dk g(k) \cos(kx) + i \int_{-\infty}^{\infty} dk g(k) \sin(kx).$$

Thus, $g(k)$ can be expressed in terms of the cosine and sine transforms of $f(x)$ as

$$g(k) = g^c(k) - ig^s(k)$$

where,

$$g^c(k) = \frac{1}{2\pi} \int_{-\infty}^{\infty} dx f(x) \cos(kx)$$

and

$$g^s(k) = \frac{1}{2\pi} \int_{-\infty}^{\infty} dx f(x) \sin(kx).$$

Similarly, $f(x)$ can also be expressed in terms of the cosine and sine inverse transforms of $g(k)$ as

$$f(x) = f^c(x) + if^s(x)$$

where

$$f^c(x) = \int_{-\infty}^{\infty} dk g(k) \cos(kx)$$

and

$$f^s(x) = \int_{-\infty}^{\infty} dk g(k) \sin(kx).$$

This representation is particularly useful for functions with odd/even symmetry. When $f(x)$ is even,

$$f(x) = f(-x),$$

the Fourier transform picks up only the even Fourier modes. Consequently, the Fourier transform is also an even function given in terms of cosine,

$$g(k) = g^c(k)$$
$$= \frac{1}{\pi} \int_0^\infty dx f(x) \cos(kx)$$

and the sine transform vanishes because the integrand is odd. Thus, the cosine transforms are

$$f(x) = f^c(x) = 2 \int_0^\infty dk g(k) \cos(kx), \qquad g(k) = g^c(k) = \frac{1}{\pi} \int_0^\infty dx f(x) \cos(kx).$$

Example 3 Determine the Fourier transform of the *non-periodic* $f(x)$ defined as

$$f(x) = \begin{cases} 1, & |x| < 1 \\ 0, & |x| > 1 \end{cases}$$

Solution: We notice that the function is even in $[-1, 1]$ where it is non-zero. The function also bounded, thus satisfies Dirichlet's conditions. Hence, we can use the cosine transform

$$g(k) = \frac{1}{\pi} \int_0^\infty dx f(x) \cos(kx)$$
$$= \frac{1}{\pi} \int_0^1 dx \cos(kx)$$
$$= \frac{\sin(k)}{k\pi}.$$

Hence,

$$f(x) = 2 \int_0^\infty dk g(k) \cos(kx)$$
$$= \frac{2}{\pi} \int_0^\infty dk \frac{\sin(k) \cos(kx)}{k}.$$

This expression allows us to find a closed form of the integral for a given x. For instance,

1 Lecture 21: Fourier Transform

$$\int_0^\infty dk \frac{\sin(k)}{k} = \frac{\pi}{2} f(0) = \frac{\pi}{2}.$$

Similarly, when $f(x)$ is *odd*

$$f(x) = f(-x),$$

the Fourier transforms pick up only the odd terms and reduce to sine transforms,

$$f(x) = 2 \int_0^\infty dk g^s(k) \sin(kx), \quad g^s(k) = \frac{1}{\pi} \int_0^\infty dx f(x) \sin(kx).$$

Example 4 Let us determine the Fourier transform of the non-periodic $f(x)$ defined as

$$f(x) = \begin{cases} x, & |x| < 1 \\ 0, & |x| > 1 \end{cases}$$

Solution: The function is odd in the interval $[-1, 1]$ and zero everywhere else. It is bounded, thus satisfies Dirichlet's conditions. Hence, we can use the Fourier sine transform

$$g^s(k) = \frac{1}{\pi} \int_0^\infty dx f(x) \sin(kx)$$

$$= \frac{1}{\pi} \int_0^1 dx\, x \sin(kx)$$

$$= -\frac{\cos(k)}{k\pi} + \frac{\sin(k)}{k^2 \pi}.$$

Thus,

$$f(x) = 2 \int_0^\infty dk g^s(k) \sin(kx)$$

$$= \frac{2}{\pi} \int_0^\infty dk \sin(kx) \left(-\frac{\cos(k)}{k} + \frac{\sin(k)}{k^2} \right).$$

Similar to the previous example, we can use this expression to find close forms of the integral for particular values of x.

1.1.3 Parseval's Relation

Theorem 2 (Parseval's theorem) *A pair of Fourier transforms*

$$f(x) = \int_{-\infty}^{\infty} dk\, g(k) e^{ikx}, \qquad g(k) = \frac{1}{2\pi} \int_{-\infty}^{\infty} dx\, f(x) e^{-ikx}$$

satisfy the completeness relation

$$\int_{-\infty}^{\infty} dx |f(x)|^2 = 2\pi \int_{-\infty}^{\infty} dk |g(k)|^2$$

To show this, we evaluate the integral over $|f(x)|^2$,

$$\int_{-\infty}^{\infty} dx |f(x)|^2 = \int_{-\infty}^{\infty} dx\, \bar{f}(x) f(x)$$

$$= \int_{-\infty}^{\infty} dx\, \bar{f}(x) \left[\int_{-\infty}^{\infty} dk\, g(k) e^{ikx} \right]$$

$$= \int_{-\infty}^{\infty} dk\, g(k) \left[\int_{-\infty}^{\infty} dx\, \bar{f}(x) e^{ikx} \right]$$

$$= 2\pi \int_{-\infty}^{\infty} dk\, g(k) \bar{g}(k)$$

$$= 2\pi \int_{-\infty}^{\infty} dk |g(k)|^2.$$

2 Lecture 22: Laplace Transform

In this lecture, we focus on the Laplace transform and some of its properties.

2.1 Laplace Transform

Theorem 3 *Let $f(t)$ be a function that is piecewise continuous on every finite interval in the range $t \geq 0$ and bounded by an exponential*

$$|f(t)| \leq M e^{\gamma t}, \text{ for all } t \geq 0$$

for same constants γ and M. Then, there exits a unique Laplace transform *of $f(t)$*

$$\boxed{F(s) = \mathcal{L}[f(t)] = \int_0^\infty dt f(t) e^{-st}}$$

This integral is finite when the integrand decays sufficiently fast at infinity. This condition determines the domain of s. Using the triangle inequality and the upper bound of $f(t)$, we have that

$$\left| \int_0^\infty dt f(t) e^{-st} \right| \leq \int_0^\infty dt |f(t)| e^{-st} \leq M \int_0^\infty dt e^{-(s-\gamma)t},$$

which is finite when $s > \gamma$.

The function $f(t)$ is also the *inverse Laplace transform* of $F(s)$ and determined formally as

$$\boxed{f(t) = \mathcal{L}^{-1}[F(s)]}$$

In this lecture, we discuss the properties of the Laplace transform. A brief discussion on the integral form of inverse Laplace transform is added in the supplementary section.

Example 5 The Laplace transform of a constant $f(t) = 1$ is

$$\mathcal{L}[1] = \int_0^\infty dt e^{-st}$$
$$= \frac{1}{s}, \quad s \geq 0$$

Example 6 The Laplace transform of the exponential $f(t) = e^{at}$ is

$$\mathcal{L}\left[e^{at}\right] = \int_0^\infty dt \, e^{-(s-a)t}$$

$$= \frac{1}{s-a}, \quad s > a.$$

Linearity of the Laplace Transform: Like the Fourier transform, the Laplace transform is a linear operation and thus satisfies the additivity property

$$\mathcal{L}[af(t) + bg(t)] = a\mathcal{L}[f(t)] + b\mathcal{L}[g(t)]$$

This is useful to Laplace transforms indirectly as exemplified below.

Example 7 The Laplace transform (LT) of

$$f(t) = \cos(at) = \frac{1}{2}\left(e^{iat} + e^{-iat}\right)$$

can be evaluated from the LT of the exponential using the additivity property,

$$\mathcal{L}[\cos(at)] = \frac{1}{2}\mathcal{L}\left[e^{iat}\right] + \frac{1}{2}\mathcal{L}\left[e^{-iat}\right]$$

$$= \frac{1}{2}\frac{1}{s - ia} + \frac{1}{2}\frac{1}{s + ia}$$

$$= \frac{s}{s^2 + a^2}, \quad s > 0$$

Example 8 When the Laplace transform is a fraction

$$F(s) = \frac{1}{(s-a)(s-b)}, \quad a \neq b$$

we can use the partial fraction decomposition to reduce it to simple fractions

$$\frac{1}{(s-a)(s-b)} = \frac{1}{a-b}\left[\frac{1}{s-a} - \frac{1}{s-b}\right]$$

to determine the inverse Laplace transform using the additive property as

$$f(t) = \mathcal{L}^{-1}\left[\frac{1}{(s-a)(s-b)}\right]$$
$$= \frac{1}{a-b}\left(\mathcal{L}^{-1}\left[\frac{1}{s-a}\right] - \mathcal{L}^{-1}\left[\frac{1}{s-b}\right]\right)$$
$$= \frac{1}{a-b}\left(e^{at} - e^{bt}\right)$$

2.1.1 Laplace Transform of Derivatives

Let $f(t)$ and $\mathcal{L}[f(t)]$ be a pair of Laplace transforms. The Laplace transform of the derivative $f'(t)$ is determined by the Laplace transform of $f(t)$ by the following transformation

$$\mathcal{L}[f'(t)] = s\mathcal{L}[f(t)] - f(0).$$

For the second derivative,

$$\mathcal{L}[f''(t)] = s^2\mathcal{L}[f(t)] - sf(0) - f'(0),$$

and for the n'th derivative

$$\boxed{\mathcal{L}[f^{(n)}(t)] = s^n\mathcal{L}[f(t)] - s^{n-1}f(0) - s^{n-2}f'(0) - \cdots f^{(n-1)}(0)} \qquad (4)$$

To show this, we may evaluate the LT of each derivative by successive integrations by parts,

$$\mathcal{L}[f'(t)] = \int_0^\infty dt\, e^{-st} f'(t) = f(t)e^{-st}\Big|_0^\infty + s\int_0^\infty dt\, e^{-st} f(t)$$
$$= -f(0) + s\mathcal{L}[f(t)]$$

$$\mathcal{L}[f''(t)] = \int_0^\infty dt\, e^{-st} f''(t) = f'(t)e^{-st}\Big|_0^\infty + s\int_0^\infty dt\, e^{-st} f'(t)$$
$$= -f'(0) + s\mathcal{L}[f'(t)]$$
$$= -f'(0) - sf(0) + s^2\mathcal{L}[f(t)]$$

$$\mathcal{L}[f^{(n)}(t)] = \int_0^\infty dt\, e^{-st} f^{(n)}(t) = f^{(n-1)}(t)e^{-st}\Big|_0^\infty + s\int_0^\infty dt\, e^{-st} f^{(n-1)}(t)$$
$$= -f^{(n-1)}(0) + s\mathcal{L}[f^{(n-1)}(t)]$$
$$= -f^{(n-1)}(0) - sf^{(n-2)}(0) \cdots - s^{n-1}f(0) + s^n\mathcal{L}[f(t)]$$

2.1.2 Derivative of Laplace Transform

Let us assume that $f(t)$ has a Laplace transform $F(s)$. Then the derivative of its transform, $F'(s)$, can be found by differentiating with respect to s inside the integral and relates to the LT of $tf(t)$

$$F'[s] = \frac{d}{ds} \int_0^\infty dt f(t) e^{-st} = -\int_0^\infty dt [tf(t)] e^{-st}$$

Thus, the useful formula

$$\boxed{\mathcal{L}[tf(t)] = -F'(s)}$$

correspondting to its inverse transform

$$\boxed{\mathcal{L}^{-1}[F'(s)] = -tf(t)}$$

Example 9 To illustrate this point, let us consider

$$f(t) = \sin(\omega t) = \frac{1}{2i}(e^{i\omega t} - e^{-i\omega t}).$$

Its Laplace transform is computed as follows

$$F(s) = \mathcal{L}[\sin \omega t] = \frac{1}{2i} \int_0^\infty ds (e^{i\omega t} - e^{-i\omega t}) e^{-st}$$

$$= \frac{1}{2i} \left(\frac{1}{s - i\omega} - \frac{1}{s + i\omega} \right)$$

$$= \frac{\omega}{s^2 + \omega^2}$$

Thus, its derivative is

$$F'(s) = -\mathcal{L}[t \sin \omega t] = \frac{d}{ds} \frac{\omega}{s^2 + \omega^2}$$

$$= -\frac{2\omega s}{(s^2 + \omega^2)^2}$$

from which we can quickly evaluate the inverse Laplace transform

$$\mathcal{L}^{-1}\left[\frac{s}{(s^2 + \omega^2)^2} \right] = \frac{t}{2\omega} \sin \omega t.$$

2 Lecture 22: Laplace Transform

Notice that the actual integral might be very difficult to solve directly. However, using the properties of the Laplace transforms, we are often able to relate them to simpler forms or even evaluate them indirectly.

Example 10 Find the inverse Laplace transform

$$\mathcal{L}^{-1}\left[\frac{4}{(s+1)^2}\right]$$

First, we notice that the Laplace transform is the derivative of this Laplace transform:

$$\frac{4}{(s+1)^2} = F'(s) \Rightarrow F(s) = -\frac{4}{s+1}.$$

Thus, the inverse Laplace transform is given by

$$\mathcal{L}^{-1}\left[-\frac{4}{s+1}\right] = -4\mathcal{L}^{-1}\left[\frac{1}{s+1}\right] = -4e^{-t},$$

such that

$$\mathcal{L}^{-1}\left[\frac{4}{(s+1)^2}\right] = 4te^{-t}.$$

2.1.3 Shifting Theorems

The shifting theorems are useful to determine how the function with shifted coordinates in the s-space or t-space transforms in the corresponding reciprocal space.

Theorem 4 (s-shift theorem) *If $f(t)$ has the Laplace transform $F(s)$ where $s > \gamma$, then*

$$\boxed{\mathcal{L}\left[e^{at}f(t)\right] = F(s-a), \quad F(s) = \mathcal{L}[f(t)]}$$

This is often used in finding the inverse Laplace transform as

$$\mathcal{L}^{-1}[F(s-a)] = e^{at}\mathcal{L}^{-1}[F(s)] = e^{at}f(t).$$

To show this, let us reevaluate the integral associated with the s-shirt $s - a$,

$$F(s-a) = \int_0^\infty dt\, e^{-(s-a)t} f(t)$$
$$= \int_0^\infty dt\, \left[e^{at} f(t)\right] e^{-st}$$
$$= \mathcal{L}\left[e^{at} f(t)\right],$$

thus, by re-arranging the integrand, we see that it be expressed as the Laplace transform of $f(t)e^{at}$.

Theorem 5 (t-shift theorem) *If $f(t)$ has the Laplace transform $F(s)$ where $s > \gamma$, then*

$$\boxed{\mathcal{L}[f(t-a)H(t-a)] = e^{-as}F(s), \qquad F(s) = \mathcal{L}[f(t)]}$$

and thus the inverse Laplace transform is

$$\mathcal{L}^{-1}\left[e^{-as}F(s)\right] = f(t-a)H(t-a)$$

where the Heaviside function is

$$H(t-a) = \begin{cases} 0, & t < a \\ 1, & t > a \end{cases}$$

To show this, we evaluate the integral corresponding to the Laplace transform of $f(t)$,

$$e^{-as}F(s) = e^{-as}\int_0^\infty d\tau\, e^{-s\tau}f(\tau).$$

Using the variable change $t = \tau + a$, we rewrite the integral as

$$e^{-as}F(s) = \int_a^\infty dt\, f(t-a)e^{-st}$$
$$= \int_0^\infty dt\, H(t-a)f(t-a)e^{-st}$$
$$= \mathcal{L}[H(t-a)f(t-a)].$$

Example 11 Using the t-shift theorem, we can quickly evaluate the Laplace transform of the Heaviside function as

$$\mathcal{L}[H(t-a)] = e^{-as}\mathcal{L}[1] = \frac{e^{-as}}{s}$$

Example 12 Let us find the inverse Laplace transform $\mathcal{L}^{-1}\left[e^{-3s}\frac{1}{s^3}\right]$.

We first find the inverse Laplace transform of $F(s) = s^{-3}$, using the derivatives of Laplace transforms

$$f(t) = \mathcal{L}^{-1}\left[\frac{1}{s^3}\right]$$
$$= -\frac{1}{2}t\mathcal{L}^{-1}\left[\frac{1}{s^2}\right]$$
$$= \frac{1}{2}t^2\mathcal{L}^{-1}\left[\frac{1}{s}\right]$$
$$= \frac{1}{2}t^2.$$

Now, we can apply the t-shift theorem for $a = 3$, and obtain that

$$\mathcal{L}^{-1}\left[e^{-3s}\frac{1}{s^3}\right] = H(t-3)f(t-3) = \frac{1}{2}(t-3)^2 H(t-3)$$

2.1.4 Applications to Initial Value Problems

Laplace transform is often used in solving linear differential equations with initial conditions. To exemplify this method, we consider a generic 2nd order ODE with non-homogeneous initial conditions given as

$$y'' + ay' + by = r(t), \quad y(0) = K_0, \quad y'(0) = K_1. \tag{5}$$

Step 1 is to apply the Laplace transform to the differential equation to find the equation satisfied by the LT of $y(t)$,

$$Y(s) = \mathcal{L}[y(t)].$$

Let us also denote the LT of the forcing term as

$$R(s) = \mathcal{L}[r(t)].$$

Using the general formula for the LT of the derivatives in Eq. 4, we find that Eq. 5 reduces to

$$[s^2 Y(s) - sy(0) - y'(0)] + a[sY(s) - y(0)] + bY(s) = R(s),$$

which can be rewritten as

$$Y(s)(s^2 + as + b) = (s+a)K_0 + K_1 + R(s).$$

Step 2 is to solve for $Y(s)$. In our case, this leads to

$$Y(s) = [(s+a)K_0 + K_1]Q(s) + R(s)Q(s), \qquad Q(s) = \frac{1}{s^2 + as + b}.$$

Step 3 is to apply the inverse LT to find $y(t)$ by inverse Laplace transform as

$$y(t) = \mathcal{L}^{-1}[Y(s)].$$

We may not easily perform the integral directly. However, we may be able to evaluate it indirectly using the shifting theorems and the properties of the Laplace transforms.

Example 13 Let us find the particular solution of this initial value problem

$$y'' - y = t, \qquad y(0) = y'(0) = 1$$

Step 1 We compute the Laplace transform of the forcing term

$$\mathcal{L}[t] = \int_0^\infty dt\, t e^{-st} = \frac{1}{s^2}$$

and apply the Laplace transform to the ode to reduces to

$$s^2 Y - 1 - s - Y = \frac{1}{s^2} \rightarrow (s^2 - 1)Y = s + 1 + \frac{1}{s^2}.$$

Step 2 We solve for $Y(s)$ given by

$$Y(s) = \frac{s+1}{s^2 - 1} + \frac{1}{s^2(s^2 - 1)}$$

$$= \frac{1}{s-1} + \frac{1}{s^2 - 1} - \frac{1}{s^2}.$$

Step 3 To find $y(t)$, we apply the inverse place transform to each term in $Y(s)$ and use the properties of the Laplace transforms

$$\mathcal{L}^{-1}\left[\frac{1}{s-1}\right] = e^t.$$

Using the partial fraction decomposition and additive properties, we find that

$$\mathcal{L}^{-1}\left[\frac{1}{s^2 - 1}\right] = \frac{1}{2}\mathcal{L}^{-1}\left[\frac{1}{s-1}\right] - \frac{1}{2}\mathcal{L}^{-1}\left[\frac{1}{s+1}\right] = \sinh t$$

The last term has the inverse LT given by

$$\mathcal{L}^{-1}\left[\frac{1}{s^2}\right] = t.$$

Thus, adding up these three inverse LTs, we find the solution as

$$y(t) = \mathcal{L}^{-1}[Y(s)] = e^t + \sinh t - t.$$

2.2 Inverse Laplace Transform Integral (Supplementary)

You may wonder what is the integral representation of the inverse Laplace transform. Recall that for the LT integral we used the upper bound of $f(t)$ to determine the domain in the s-space where the LT integral is finite, i.e. $F(s)$ is well-defined. $F(s)$ is intimately related to the Fourier transform when we let s be a complex number. It turns out that the integral for the inverse LT needs to be extended into the complex s plane to recover a well-defined $f(t)$ for $t > 0$.

To see this, let us start from a complex variable $s = s_1 + i s_2$ and write the Laplace transform integral in these coordinates

$$F(s) = \int_0^\infty dt f(t) e^{-s_1 t} e^{-i s_2 t},$$

which is well-defined when $s_1 \equiv Re(s) > \gamma$. This integral can be extended to the whole axis by using the Heaviside function $H(t)$ in the integrand

$$F(s) = \int_{-\infty}^\infty dt \left[H(t) f(t) e^{-s_1 t} \right] e^{-i s_2 t}.$$

We notice that this is the Fourier transform of $H(t) f(t) e^{-s_1 t}$. Thus, we can evaluate this function by the inverse Fourier transform integral

$$H(t) f(t) e^{-s_1 t} = \frac{1}{2\pi} \int_{-\infty}^\infty ds_2 F(s_1 + i s_2) e^{i s_2 t}.$$

For $t > 0$, the Heaviside function equals unity, therefore it follows that the function $f(t)$ has the integral representation

$$f(t) = \frac{1}{2\pi} \int_{-\infty}^\infty ds_2 F(s_1 + i s_2) e^{(s_1 + i s_2) t},$$

which is the line integral in the complex plane (s_1, s_2) along a line parallel to the imaginary axis, i.e. for some constant $s_1 = c$. Using the line parameterization, $s = c + is_2$ with $ds = ids_2$, we re-write the integral as a complex line integral representing the *inverse Laplace transform*

$$f(t) = \frac{1}{2\pi i} \int_\Gamma ds\, F(s) e^{st},$$

where Γ is the vertical line $s_1 = c > \gamma$ in the (s_1, s_2) complex plane. Often this is equivalently represented by the integration limits

$$f(t) = \frac{1}{2\pi i} \int_{c-i\infty}^{c+i\infty} ds\, F(s) e^{st}.$$

This is also known as the *Bromwich integral* and is typically evaluated using complex analysis.

3 Lecture 23: Convolution of Integral Transforms

In this lecture, we focus on applications of the Fourier and Laplace transforms to solving ordinary differential equations. We learn how these integral transforms are intimately connected with the Green's functions and convolution integrals.

3.1 *Integral Transform of a Convolution*

The solution of a linear, non-homogeneous differential equation follows as a convolution integral of the Green's function with the sourcing term. Using the integral transforms, we can determine the Green's function in the conjugate space (k-space for FT or s-space of LT) where the corresponding convolution integral takes a simpler form. Before we take concrete examples, let us see how the convolution integral transforms under Fourier or Laplace transformation.

Theorem 6 (Convolution Theorem for Fourier transforms) *Let us consider two pairs of* Fourier transforms

$$F(k) = \mathcal{F}[f(x)] = \frac{1}{2\pi} \int_{-\infty}^{\infty} dx\, f(x) e^{-ikx}, \quad f(x) = \mathcal{F}^{-1}[F(k)] = \int_{-\infty}^{\infty} dk\, F(k) e^{ikx}$$

$$G(k) = \mathcal{F}[g(x)] = \frac{1}{2\pi} \int_{-\infty}^{\infty} dx\, g(x) e^{-ikx}, \quad g(x) = \mathcal{F}^{-1}[G(k)] = \int_{-\infty}^{\infty} dk\, G(k) e^{ikx}$$

3 Lecture 23: Convolution of Integral Transforms

Given the product function $H(k) = F(k) \cdot G(k)$ in the k-space, there exits a function $h(x)$ which is the convolution *of $f(x)$ and $g(x)$ defined as:*

$$h(x) \equiv \mathcal{F}^{-1}[2\pi H(k)] \equiv (f * g)(x) = \int_{-\infty}^{\infty} du\, f(u) g(x-u)$$

Equivalently,

$$\mathcal{F}[h(x)] = \mathcal{F}[f(x) * g(x)] = 2\pi H(k) = 2\pi F(k) \cdot G(k)$$

where we use the "$*$" symbol as short-hand notation of the convolution integral.

We know that in the Fourier space, $H(k)$ is the product of the Fourier transforms:

$$H(k) = F(k) \cdot G(k)$$
$$= \left(\frac{1}{2\pi} \int_{-\infty}^{\infty} du\, e^{-iku} f(u)\right) \cdot \left(\frac{1}{2\pi} \int_{-\infty}^{\infty} dw\, e^{-ikw} g(w)\right)$$
$$= \frac{1}{(2\pi)^2} \int_{-\infty}^{\infty} \int_{-\infty}^{\infty} du\, dw\, e^{-ik(u+w)} f(u) g(w).$$

In the integral over w, we can use the change of variable $x = w + u$ such that

$$H(k) = \frac{1}{(2\pi)^2} \int_{-\infty}^{\infty} \int_{-\infty}^{\infty} dx\, du\, e^{-ikx} f(u) g(x-u)$$
$$= \frac{1}{(2\pi)^2} \int_{-\infty}^{\infty} dx\, e^{-ikx} \int_{-\infty}^{\infty} du\, f(u) g(x-u)$$
$$= \frac{1}{2\pi} \mathcal{F}[h(x)].$$

Application to Boundary Value Problem: The Fourier transform is an elegant method of solving linear differential equations under homogeneous boundary conditions. Let us consider as an example the generic equation for a forced and damped harmonic oscillator given by

$$y'' + \gamma y' + \omega^2 y = f(x),$$

where $\gamma > 0$ is the constant damping rate, ω is the internal frequency of the oscillator and $f(x)$ is some external driving force. We are seeking out the solution $y(x)$ which vanishes sufficiently fast at $\pm\infty$. Thus, the function satisfies the Dirichlet boundary conditions. The Fourier transform method is applied in three main steps:

Step 1 is to apply the Fourier transform on the differential equation and find the corresponding algebraic equation satisfied by $\hat{y}(k) = \mathcal{F}[f(x)]$. In our case, this results in

$$\mathcal{F}[y''(x) + \gamma y'(x) + \omega^2 y(x)] = \hat{f}(k)$$

where $\hat{f}(k) = \mathcal{F}[f(x)]$ is the FT of the source term. FT is a linear transformation just like the Laplace transform. Thus, using the additive property when γ, ω are constants, the Fourier transform can be applied on each term on the left hand side. This leads to

$$(-k^2 + i\gamma k + \omega)\hat{y}(k) = \hat{f}(k).$$

Step 2 is to solve for $\hat{y}(k)$. This implies that

$$\hat{y}(k) = \frac{1}{-k^2 + i\gamma k + \omega} \mathcal{F}[f(x)]$$
$$= 2\pi \hat{\chi}(k) \hat{f}(k)$$

Notice that the solution in the Fourier space is a product of two Fourier transforms. One is the external force $\hat{f}(k)$. The other one

$$\hat{\chi}(k) = (2\pi)^{-1}(-k^2 + i\gamma k + \omega)^{-1},$$

which is the *response function* that tells us how the oscillator responds to an external perturbation.

Step 3 is to apply the inverse Fourier transform and obtain the solution in the position space,

$$y(x) = 2\pi \int_{-\infty}^{\infty} dk \, \hat{\chi}(k) \hat{f}(k) e^{ikx}. \tag{6}$$

Equivalently, we can write this as a convolution integral in the position space using $\hat{f}(k) = \frac{1}{2\pi} \int_{-\infty}^{\infty} dx' f(x') e^{-ikx'}$, thus

$$y(x) = \int_{-\infty}^{\infty} dx' \chi(x - x') f(x')$$

where $\chi(x) = \int_{-\infty}^{\infty} dk \, \hat{\chi}(k) e^{ikx}$ is the response function in the position space. Thus, we have two routes for finding the solution and can choose which integral is most convenient to solve.

Theorem 7 (Convolution Theorem of Laplace transforms) *Consider two functions with* Laplace transforms

$$F(s) = \mathcal{L}[f(t)], \qquad f(t) = \mathcal{L}^{-1}[F(s)]$$

3 Lecture 23: Convolution of Integral Transforms

$$G(s) = \mathcal{L}[g(t)], \quad g(t) = \mathcal{L}^{-1}[G(s)]$$

Given $H(s) = F(s)G(s)$, there exits a function $h(t)$ which is the convolution of $f(t)$ and $g(t)$ defined as:

$$h(t) \equiv \mathcal{L}^{-1}[H(s)] \equiv (f * g)(t) = \int_0^t d\tau f(\tau) g(t - \tau)$$

where $*$ is a shorthand operator notation for the convolution.

We use the t-shifting theorem to find an integral expression

$$e^{-s\tau} G(s) = \mathcal{L}[g(t-\tau) H(t-\tau)]$$
$$= \int_0^\infty dt\, e^{-st} g(t-\tau) H(t-\tau).$$

We use this integral form for $e^{-s\tau}G(s)$ when evaluating the product of the two Laplace transforms:

$$H(s) = F(s) \cdot G(s)$$
$$= \left(\int_0^\infty d\tau\, e^{-s\tau} f(\tau)\right) G(s) = \int_0^\infty d\tau f(\tau) \left(e^{-s\tau} G(s)\right)$$
$$= \int_0^\infty d\tau f(\tau) \left(\int_0^\infty dt\, e^{-st} g(t-\tau) H(t-\tau)\right).$$

Changing the order of integration, we have that

$$H(s) = \int_0^\infty dt\, e^{-st} \left(\int_0^\infty d\tau\, H(t-\tau) f(\tau) g(t-\tau)\right)$$
$$= \int_0^\infty dt\, e^{-st} \left(\int_0^t d\tau f(\tau) g(t-\tau)\right)$$
$$= \int_0^\infty dt\, e^{-st} h(t) = \mathcal{L}[h(t)].$$

Thus,

$$h(t) = \mathcal{L}^{-1}[G(s)] = \int_0^t d\tau f(\tau) \cdot g(t-\tau).$$

Note that $f * g \equiv g * f$, thus we can interchange the arguments τ and $t - \tau$ between the two functions.

Example 14 Given
$$\mathcal{L}^{-1}\left[\frac{1}{s^2 + 1}\right] = \sin(t)$$

we want to find the inverse Laplace transform of $\left[\frac{1}{(s^2+1)^2}\right]$. By the convolution theorem for the product of the Laplace transforms, it follows that the inverse LT is the convolution of the two sine functions

$$\mathcal{L}^{-1}\left[\frac{1}{(s^2 + 1)^2}\right] = \mathcal{L}^{-1}\left[\frac{1}{s^2 + 1}\right] * \mathcal{L}^{-1}\left[\frac{1}{s^2 + 1}\right]$$
$$= \int_0^t d\tau \sin(\tau) \sin(t - \tau)$$
$$= -\frac{1}{4} \int_0^t d\tau (e^{i\tau} - e^{-i\tau})(e^{i(t-\tau)} - e^{-i(t-\tau)})$$
$$= -\frac{1}{4} \int_0^t d\tau (e^{it} + e^{-it} - e^{i(t-2\tau)} - e^{-i(t-2\tau)})$$
$$= -\frac{1}{4} \int_0^t d\tau (2 \cos t - 2 \cos(t - 2\tau))$$
$$= \frac{1}{2} t \cos t + \frac{1}{2} \sin(t).$$

Application to Initial Value Problem: Let us consider the same damped harmonic oscillator but this time with the homogeneous initial conditions

$$y''(t) + \gamma y'(t) + \omega^2 y(t) = r(t), \quad y(0) = y'(0) = 0$$

which has the solution given as an inverse Laplace transform $y(t) = \mathcal{L}^{-1}[Y(s)]$. The Laplace transform function $Y(s)$ is obtained by the LT of the ode leading to

$$Y(s) = R(s) Q(s),$$

with $R(s) = \mathcal{L}[r(t)]$ being the LT of the forcing term, and

$$Q(s) = \frac{1}{s^2 + \gamma s + \omega^2}$$

which represents the Laplace transform of the Green's function denoted as $q(t) = \mathcal{L}^{-1}[Q(s)]$. By the convolution theorem, the solution can be expressed as the convolution of the Green's function with the forcing term

$$y(t) = \mathcal{L}^{-1}[Y(s)] \tag{7}$$

$$= \int_0^t d\tau q(\tau) r(t-\tau). \tag{8}$$

3.2 Boundary Value Problem

Example 15 Let us consider the following damped and forced harmonic oscillator

$$y'' + 2y' + y = \delta(x),$$

where the force term is the Dirac delta function which reduces to a constant value in Fourier space,

$$\mathcal{F}[\delta(x)] = \frac{1}{2\pi} \int_{-\infty}^{\infty} dx \delta(x) e^{-ikx} = \frac{1}{2\pi}.$$

Thus, the solution from Eq. 6 is given by

$$y(x) = \frac{1}{2\pi} \int_{-\infty}^{\infty} dk \frac{1}{-k^2 + 2ik + 1} e^{ikx}$$

$$= -\frac{1}{2\pi} \int_{-\infty}^{\infty} dk \frac{e^{ikx}}{(k-i)^2}$$

$$= -\frac{1}{2\pi} I$$

We can solve this I integral using complex analysis. For $x > 0$, the integral equals the contour integral in the upper half plane where e^{izx} is an exponentially decaying function. The integrand has a pole of order 2 at $z = i$, and by the residue theorem, the contour integral is

$$x > 0 : I = \oint_{C_+} dz \frac{e^{izx}}{(z-i)^2}$$

$$= 2\pi i \operatorname{Res}\left[\frac{e^{izx}}{(z-i)^2}, z = i\right]$$

$$= 2\pi i (ixe^{-x}) = -2\pi x e^{-x}$$

For $x < 0$, e^{izx} is exponentially decaying ($e^{-xIm(z)}$) in the lower half-plane where the integrand is analytic. Thus, by the Cauchy's theorem

$$x \geq 0 : I = \oint_{C_-} dz \frac{e^{izx}}{(z-i)^2} = 0$$

Thus, the solution is

$$y(x) = \begin{cases} xe^{-x}, & x > 0 \\ 0, & x \leq 0 \end{cases}$$

Example 16 Let us now take a similar equation but with a different sourcing term

$$y''(x) + 2y'(x) + y(x) = e^{-x} H(x).$$

The Fourier transform of this equation leads to

$$(-k^2 + 2ik + 1)\hat{y}(k) = \mathcal{F}[e^{-x} H(x)]$$

which implies that the FT of the Green function is

$$\hat{\chi}(k) = \frac{1}{2\pi} \frac{1}{-k^2 + 2ik + 1} = \frac{1}{2\pi} \frac{1}{(ik+1)^2} = -\frac{1}{2\pi} \frac{1}{(k-i)^2}$$

and the solution in the k space is

$$\hat{y}(k) = 2\pi \hat{\chi}(k) \mathcal{F}[e^{-x} H(x)]$$

We evaluate the Green function in the position space as

$$\chi(x) = \mathcal{F}^{-1}[\hat{\chi}(k)] \tag{9}$$

$$= -\frac{1}{2\pi} \int_{-\infty}^{\infty} dk \frac{e^{ikx}}{(k-i)^2} = xe^{-x} H(x). \tag{10}$$

Since both the forcing term and the Green's function are non-zero for $x > 0$, the solution will also be non-zero for $x > 0$. By the convolution theorem, the solution for $x > 0$ is

$$y(x) = \int_{-\infty}^{\infty} du \, \chi(x-u) e^{-u} H(u)$$

$$= \int_{0}^{\infty} du (x-u) e^{-u} e^{-x+u} H(x-u) = \int_{0}^{x} du (x-u) e^{-u} e^{-x+u}$$

$$= e^{-x} \int_{0}^{x} du (x-u) = \frac{1}{2} x^2 e^{-x}$$

Thus, for any x:
$$y(x) = \frac{1}{2}x^2 e^{-x} H(x).$$

3.3 Initial Value Problem

Let us now take few concrete examples of odes with initial conditions where we apply the Laplace transform method.

Example 17 Let us consider this ode
$$y'' - y = H(t - a),$$
with homogeneous initial condition $y(0) = y'(0) = 0$. The Heaviside function is
$$H(t - a) = \begin{cases} 1, & t > a \\ 0, & t < a \end{cases}.$$

Recall from the previous lecture that the Laplace transform of the Heaviside function is
$$\begin{aligned} R(s) &= \mathcal{L}[H(t - a)] \\ &= \int_0^\infty H(t - a) e^{-st} \\ &= \frac{e^{-as}}{s}. \end{aligned}$$

Inserting this into the Laplace transform of the ode and using the additive property of the LT, we find that the Laplace transform of the solution $Y(s) = \mathcal{L}[y(t)]$ is given by
$$(s^2 - 1)Y(s) = \frac{e^{-as}}{s} \Rightarrow Y(s) = \frac{e^{-as}}{s(s^2 - 1)}.$$

By partial fraction decomposition, we can rewrite it equivalently as
$$Y(s) = e^{-as}\left(\frac{1}{2(s+1)} + \frac{1}{(s-1)} - \frac{1}{s}\right) = e^{-as} F(s).$$

Using the inverse LT of the simple fractions
$$\mathcal{L}^{-1}\left[\frac{1}{s}\right] = 1, \quad \mathcal{L}^{-1}\left[\frac{1}{2}\left(\frac{1}{s+1} + \frac{1}{s-1}\right)\right] = \cosh(t),$$

we then find that the solution is

$$y(t) = \mathcal{L}^{-1}[Y(s)] = \mathcal{L}^{-1}[e^{-as}F(s)]$$

The inverse transform can be evaluated using the t-shifting theorem and leads to

$$y(t) = H(t-a)[\cosh(t-a) - 1]$$
$$= \begin{cases} 0, & t < a \\ \cosh(t-a) - 1, & t > a \end{cases}$$

Example 18 Consider the harmonic oscillator with linear damping, initially at rest and experiencing a sharp kick at $t = a$ as described by

$$y'' + 3y' + 2y = \delta(t-a), \qquad y(0) = y'(0) = 0$$

We find the solution $y(t)$ by the Laplace transform method. To find the solution, we apply the Laplace transform to the differential equation. For the Dirac delta, this is

$$\mathcal{L}[\delta(t-a)] = \int_0^\infty \delta(t-a)e^{-st} = e^{-as},$$

thus,

$$s^2 Y + 3sY + 2Y = e^{-as} \Rightarrow Y(s) = \frac{e^{-as}}{s^2 + 3s + 2} = \frac{e^{-as}}{(s+1)(s+2)}.$$

Using the partial fraction decomposition, we simplify the expression above as

$$Y(s) = e^{-as}\left(\frac{1}{s+1} - \frac{1}{s+2}\right) = e^{-as}F(s).$$

Now, we make use of the inverse LT of simple fractions

$$\mathcal{L}^{-1}\left[\frac{1}{s+1}\right] = e^{-t}, \qquad \mathcal{L}^{-1}\left[\frac{1}{s+2}\right] = e^{-2t},$$

and apply the t-shifting theorem to find the solution as

$$y(t) = \mathcal{L}^{-1}\left[e^{-as}F(s)\right] = H(t-a)\left[e^{-(t-a)} - e^{-2(t-a)}\right].$$

Equivalently, this can be written as

$$y(t) = \begin{cases} 0, & t < a \\ e^{-(t-a)} - e^{-2(t-a)}, & t > a \end{cases}$$

3 Lecture 23: Convolution of Integral Transforms

Example 19 Consider this harmonic oscillator with homogeneous initial conditions

$$y''(t) + 3y'(t) - 4y(t) = e^{3t}, \qquad y(0) = y'(0) = 0.$$

The Laplace transform satisfies that

$$(s^2 + 3s - 4)Y(s) = \mathcal{L}[e^{3t}],$$

which implies that

$$Y(s) = \frac{1}{s^2 + 3s - 4}\mathcal{L}[e^{3t}] = \frac{1}{(s+4)(s-1)}\mathcal{L}[e^{3t}]$$

We use the partial fraction decomposition to evaluate

$$\mathcal{L}^{-1}\left[\frac{1}{(s+4)(s-1)}\right] = \frac{1}{5}(e^t - e^{-4t}).$$

Thus, since $Y(s)$ is a product, we use the convolution theorem to evaluate the solution as

$$\begin{aligned} y(t) &= \frac{1}{5}\int_0^t d\tau (e^\tau - e^{-4\tau})e^{3(t-\tau)} \\ &= \frac{1}{5}e^{3t}\int_0^t d\tau (e^{-2\tau} - e^{-7\tau}) \\ &= \frac{1}{5}e^{3t}\left[\frac{5}{14} - \frac{1}{2}e^{-2t} + \frac{1}{7}e^{-7t}\right]. \end{aligned}$$

Partial Differential Equations

1 Lecture 24: Separation of Variable Method

We often use partial differential equations to describe the evolution in space and time of physical systems. We describe the transport properties of mass, momentum or energy in terms of differential equations such as the continuity equation, the Navier-Stokes equations for fluid flow, heat diffusion, wave equations, and so forth.

In this lecture, we introduce the method of separation of variables to solve linear, second-order partial differential equations with homogeneous boundary conditions.

1.1 Definitions

Let us start by considering a scalar function $u(x_1, x_2, \ldots, x_n)$ of n **independent** variables $x_1, \ldots, x_n$. Equivalently, we can collect the coordinates into a vector form $\vec{x}$ and denote the scalar field as $u(\vec{x})$.

Definition 1 A *partial derivative* of $u(\vec{x})$ with respect to the variables x_i is determined by the change in u due to incremental changes in x_i keeping all the other variables fixed,

$$\frac{\partial u}{\partial x_i} \equiv \lim_{dx_i \to 0} \frac{u(x_1, \ldots, x_i + dx_i, \ldots x_n) - u(x_1, \ldots, x_i, \ldots x_n)}{dx_i}.$$

This can be generalized to any n'th order partial derivative, with the difference that one can also have mixed terms such as

$$\frac{\partial^2 u}{\partial x_i^2}, \quad \frac{\partial^2 u}{\partial x_i \partial x_j}.$$

Definition 2 A *partial differential equation* (pde) for the function $u(\vec{x})$ is an equation that relates u with its derivatives as

$$F[u, D^1 u, D^2 u, \ldots D^n u] = g(\vec{x}),$$

where $D^n u$ is the n'th derivative with respect to any of the coordinates.

Definition 3 The *order* of a pde is determined by the highest order in derivatives.

Examples of second order pde's:

$$\text{Diffusion equation:} \quad \frac{\partial u}{\partial t} = c^2 \frac{\partial^2 u}{\partial x^2}$$

$$\text{Wave equation:} \quad \frac{\partial^2 u}{\partial t^2} = c^2 \frac{\partial^2 u}{\partial x^2}$$

Definition 4 (*Linear pde*) A *linear* pde has a normal form where the coefficients are either constants or depend *only* on the variables $\vec{x}$. The normal form of a linear pde of second order and with constant coefficients reads as

$$a_{ij} \frac{\partial^2 u}{\partial x_i \partial x_j} + b_k \frac{\partial u}{\partial x_k} + cu = g(\vec{x}),$$

where a_{ij}, b_k, c are constant coefficients. We use Einstein summation convention that repeated indices are summed over.

1.2 Separation of Variable Method

The separation of variable method is often used to solve **homogeneous and linear pde's** under appropriate boundary conditions when the dependence of u on its variables can be somehow factorized. We will present this method for functions of two variables. For transport equations, one variable is time while the other is the spatial coordinate of a one-dimensional system. This could represent the elastic vibration of a string, pressure waves in a one dimensional wire, heat diffusion in a thin bar, and so on.

1.2.1 Function of Two Variables $u(t, x)$

We consider a function $u(t, x)$ of two independent variables where t is the time variable and x is the space variable. The function satisfies a *homogeneous* linear, 2nd order pde with normal form given as

1 Lecture 24: Separation of Variable Method

$$a_{11}\frac{\partial^2 u}{\partial t^2} + 2a_{12}\frac{\partial^2 u}{\partial t \partial x} + a_{22}\frac{\partial^2 u}{\partial x^2} + b_1\frac{\partial u}{\partial t} + b_2\frac{\partial u}{\partial x} + cu = 0 \quad (1)$$

We set out to find the specific solution of this pde under **homogeneous boundary conditions** in a finite domain

$$u(x_1, t) = 0, \quad u(x_2, t) = 0, \quad \text{for every } t$$

and arbitrary initial conditions. The separation of variable method has these basic steps:

Step 1: Separate Variables We seek for a solution where the function can be written as a product of functions of a single variable as

$$u(t, x) = F(x)G(t).$$

By inserting this ansatz in Eq. 1, we can reduce the pde to a set of two ode's satisfied by $F(x)$ and $G(t)$, respectively.

Step 2: Spectrum of Independent Solutions We determine the *independent* solutions for $F(x)$ and $G(t)$ by solving the corresponding ode's. We construct the infinite set of **orthogonal functions** $\{u_n(t, x)\}$ which form a complete basis in which the specific solution can be expanded.

Step 3: General Solution We determine the general solution of the pde as a linear superposition of the independent solution $\{u_n(t, x)\}$

$$u(x, t) = \sum_n B_n u_n(x, t),$$

where the coefficients B_n are fixed by the initial conditions. We will see shortly, that B_n's represent the Fourier coefficients corresponding to the initial profile.

To exemplify this, we consider two representative homogeneous pde's, such as the *wave equation* and the *diffusion equation* in 1D.

Wave Equation: The wave equation in one dimension is a canonical example of a hyperbolic equation. It models, for instance, the elastic vibrations of a thin string. It reads as

$$\frac{\partial^2 u}{\partial t^2} = c^2 \frac{\partial^2 u}{\partial x^2},$$

where $u(x, t)$ represents the local displacement of the string relative to its equilibrium at position x and at time t. The constant c represents the speed of sound in the string. Let us go through the steps of applying the separation of variable method. Suppose we have a string of length L and that its displacement satisfies the homogeneous boundary conditions

$$u(0, t) = u(L, t) = 0 \text{ for any } t.$$

This means that the string has fixed positions at its ends. These are also called the Dirichlet's boundary conditions.

Step 1: We construct the ansatz solution as a product of two functions $u(t, x) = F(x)G(t)$ and insert it into the 1D wave equation, to obtain the following relation between the functions $F(x)$ and $G(t)$ and their derivatives

$$F''(x)G(t) = c^{-2} F(x)\ddot{G}(t) \Rightarrow \frac{F''(x)}{F(x)} = \frac{1}{c^2}\frac{\ddot{G}(t)}{G(t)},$$

where $F''(x) = \frac{d^2 F}{dx^2}$ and $\ddot{G} = \frac{d^2 G}{dt^2}$. Since these are independent functions, the two ratios are the same when they equal a constant. Let us denote the proportionality constant as $-k^2$ (this is a convenient choice as seen later) such that

$$\frac{F''(x)}{F(x)} = \frac{1}{c^2}\frac{\ddot{G}(t)}{G(t)} = -k^2$$

which means that we can reduce the wave equation to these two ode's

$$F''(x) = -k^2 F(x)$$
$$\ddot{G}(t) = -k^2 c^2 G(t). \qquad (2)$$

Step 2: We can straightforwardly find the *independent* solutions of $F(x)$ and $G(t)$ corresponding to Eq. (2), namely as

$$F_1(x) = \cos(kx), \quad F_2(x) = \sin(kx)$$

giving as the spatial vibrations of wavelengths $2\pi/k$ and

$$G_1(t) = \cos(ckt), \quad G_2(t) = \sin(ckt)$$

corresponding to temporal oscillations with frequency $\omega = ck$. The general solutions for $F(x)$ and $G(t)$ will then be linear superpositions of the corresponding independent functions with coefficients determined by the boundary conditions and initial conditions, respectively.

The boundary conditions act on the function

$$F(x) = a\sin(kx) + b\cos(kx)$$

and imply that
$$F(0) = F(L) = 0$$

and consequently that

1 Lecture 24: Separation of Variable Method

$$b = 0$$
$$a\sin(kL) + b\cos(kL) = 0 \tag{3}$$

The second equation reduces to $\sin(kL) = 0$ from which we determine that the constant k must take discrete values

$$k = n\frac{\pi}{L}, \quad n \in \mathbb{Z}$$

This also implies that we have a discrete and infinite spectrum of independent solutions for the pde. This is also known as the eigenfunction spectrum and given by

$$\left\{ \sin\left(n\frac{\pi x}{L}\right)\cos\left(n\frac{\pi ct}{L}\right), \sin\left(n\frac{\pi x}{L}\right)\sin\left(n\frac{\pi ct}{L}\right) \right\}_n.$$

Step 3: We can now construct the general solution $u(x, t)$ as a series expansion in the basis of these eigenfunctions as

$$u(x, t) = \sum_n \left[A_n \cos\left(n\frac{\pi ct}{L}\right) + B_n \sin\left(n\frac{\pi ct}{L}\right) \right] \sin\left(n\frac{\pi x}{L}\right),$$

where the coefficients A_n and B_n are determined by imposing appropriate initial conditions on $u(x, t)$ at $t = 0$. For a given initial profile $u(x, 0) = f(x)$, the general series expansion reduces to the Fourier expansion

$$f(x) = \sum_n A_n \sin\left(n\frac{\pi x}{L}\right),$$

from which it follows that the A_n are the Fourier coefficients of the sine series of the initial profile. Thus,

$$A_n = \frac{2}{L} \int_0^L dx f(x) \sin\left(n\frac{\pi x}{L}\right)$$

Since we have a second order derivatives in time, we need another initial condition which is applied on the first derivative. For an initial velocity profile

$$\left.\frac{\partial u}{\partial t}\right|_{t=0} = g(x)$$

the general series expansion reduces to a cosine expansion of $g(x)$

$$g(x) = \sum_n \frac{n\pi c}{L} B_n \sin\left(\frac{n\pi x}{L}\right) = g(x)$$

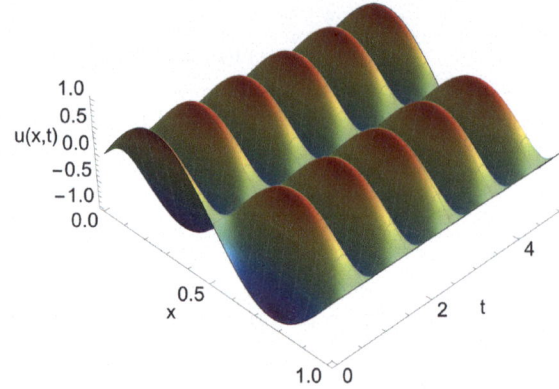

Fig. 1 Solution of the 1D wave equation for $L=1$, $c=1$ and initial conditions $f(x) = \sin(2\pi x)$ and $g(x) = 0$

where the B_n are the Fourier coefficients determined by

$$B_n = \frac{2}{n\pi c} \int_0^L dx g(x) \sin\left(\frac{n\pi x}{L}\right).$$

The solution for **homogeneous** boundary conditions and initial conditions $u(x,0) = f(x)$, $\partial_t u(x,0) = g(x)$ reads as (see Fig. 1)

$$u(x,t) = \sum_n \left[A_n \cos\left(\frac{n\pi ct}{L}\right) + B_n \sin\left(\frac{n\pi ct}{L}\right) \right] \sin\left(\frac{n\pi x}{L}\right)$$

$$A_n = \frac{2}{L} \int_0^L dx f(x) \sin\left(\frac{n\pi x}{L}\right), \quad B_n = \frac{2}{n\pi c} \int_0^L dx g(x) \sin\left(\frac{n\pi x}{L}\right).$$

Notice that the homogeneous boundary conditions imply that $f(x)$ and $g(x)$ must vanish at the boundary points.

Diffusion Equation: The diffusion equation is another important pde for transport phenomena. For a one-dimensional system, this reads as

$$\frac{\partial u}{\partial t} = c^2 \frac{\partial^2 u}{\partial x^2},$$

where $u(x,t)$ would now represent the field which spreads out in space and time with a rate set by the diffusivity coefficient c^2.

Step 1: Using the separation of variables $u(t,x) = F(x)G(t)$, we reduce the diffusion equation to this relation

1 Lecture 24: Separation of Variable Method

$$\frac{F''(x)}{F(x)} = \frac{1}{c^2}\frac{\dot{G}(t)}{G(t)} = -k^2$$

where k is some proportionality constant determined by the boundary conditions. This implies that $F(x)$ and $G(t)$ satisfy the following ode's

$$F''(x) = -k^2 F(x) \tag{4}$$
$$\dot{G}(t) = -c^2 k^2 G(t) \tag{5}$$

Step 2: The corresponding independent solutions of $F(x)$ and $G(t)$ are then

$$F_1(x) = \cos(kx), \quad F_2(x) = \sin(kx)$$

and

$$G_1(t) = e^{-c^2 k^2 t}$$

From the homogeneous boundary conditions,

$$u(0,t) = u(L,t), \text{ for any } t \Rightarrow F(0) = F(L) = 0.$$

By the same arguments as for the wave equation (the equation for $F(x)$ is the same), we find that

$$k = \frac{\pi}{L}n, \quad n \in \mathbb{Z}.$$

This implies that the independent solutions of the diffusion equation are given by the eigenfunction spectrum

$$\left\{\sin\left(n\frac{\pi x}{L}\right) e^{-\lambda_n^2 t}\right\}$$

where $\lambda_n = n\pi c/L$.

Step 3: The general solution $u(x,t)$ follows then as a series expansion in this basis

$$u(x,t) = \sum_n A_n e^{-\lambda_n^2 t} \sin(n\pi x/L)$$

with the coefficients A_n determined by the initial condition. For a given initial profile $u(x,0) = f(x)$, the expansion series reduces to the sine series of the initial profile

$$f(x) = \sum_n A_n \sin\left(n\frac{\pi x}{L}\right)$$

with the Fourier coefficients determined as

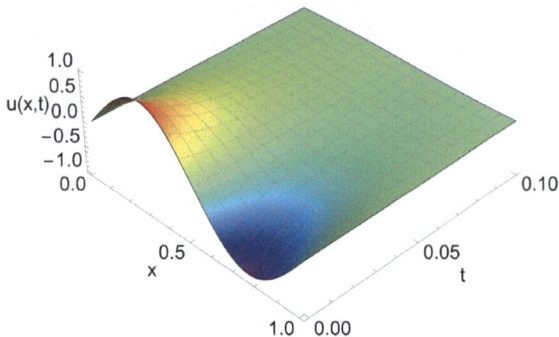

Fig. 2 Solution of the 1D diffusion equation for $L = 1$, $c = 1$ and the initial profile $f(x) = \sin(2\pi x)$

$$A_n = \frac{2}{L} \int_0^L dx f(x) \sin\left(n\frac{\pi x}{L}\right).$$

Thus, the solution for homogeneous boundary conditions and an initial condition $u(x, 0) = f(x)$ is given by (see Fig. 2)

$$u(x, t) = \sum_n A_n e^{-\lambda_n^2 t} \sin(n\pi x/L)$$

$$A_n = \frac{2}{L} \int_0^L dx f(x) \sin(n\pi x/L)$$

1.3 Diffusion Equation in 2D: Cartesian Coordinates

The separation of variables can also be applied to functions of more than two variables. We will apply this method on few specific examples and start with the heat diffusion in a rectangular spatial domain. The corresponding pde is given by

$$\frac{\partial u}{\partial t} = c^2 \left[\frac{\partial^2 u}{\partial x^2} + \frac{\partial^2 u}{\partial y^2} \right]$$

supplemented with the Dirichlet boundary conditions

$$u(x, y, t) = 0, \quad (x, y) \text{ on } [0, L_x] \times [0, L_y], \text{ for any } t \geq 0$$

and an initial profile in space given by

$$u(x, y, 0) = f(x, y).$$

1 Lecture 24: Separation of Variable Method

Step 1: By separation of variables, we write a solution as a product of three function of single variables,

$$u(t, x, y) = F(x)H(y)G(t)$$

which satisfy the following relations

$$\frac{F''(x)}{F(x)} + \frac{H''(y)}{H(y)} = \frac{1}{c^2}\frac{\dot{G}(t)}{G(t)}.$$

By the same arguments as for the 1D diffusion, it follows that

$$\frac{F''(x)}{F(x)} = -k^2, \quad \frac{H''(y)}{H(y)} = -p^2, \quad \frac{1}{c^2}\frac{\dot{G}(t)}{G(t)} = -(k^2 + p^2)$$

where k and p are constants fixed by boundary conditions.

Step 2: The independent solutions for $F(x)$, $H(y)$ are given in terms of the sine and cosine functions, just like in the 1D case. Furthermore, the homogeneous boundary conditions pick out only the sine solutions, namely

$$F''(x) + k^2 F(x) = 0, \quad \text{with } F(0) = F(L_x) = 0 \Rightarrow$$

$$F(x) = \sin\left(n\frac{\pi x}{L_x}\right), \quad k = n\frac{\pi}{L_x},$$

and similarly

$$H''(y) + p^2 F(y) = 0, \quad \text{with } H(0) = H(L_y) = 0 \Rightarrow$$

$$H(y) = \sin\left(m\frac{\pi y}{L_y}\right), \quad p = m\frac{\pi}{L_y}.$$

The corresponding equation for $G(t)$ is the same as in the 1D case

$$\dot{G}(t) + \lambda_{mn}^2 G(t) = 0 \rightarrow G(t) = e^{-\lambda_{mn}^2 t}$$

where

$$\lambda_{mn} = \frac{n^2\pi^2 c^2}{L_x^2} + \frac{n^2\pi^2 c^2}{L_y}.$$

Hence, the spectrum of independent functions for the heat diffusion in 2D is given by

$$\left\{\sin\left(n\frac{\pi x}{L_x}\right)\sin\left(m\frac{\pi y}{L_y}\right)e^{-\lambda_{mn}^2 t}\right\}$$

Step 3: General solution $u(x, t)$ is then given as

$$u(x, y, t) = \sum_n \sum_m A_{nm} e^{-\lambda_{nm}^2 t} \sin\left(n\frac{\pi x}{L_x}\right) \sin\left(m\frac{\pi y}{L_y}\right)$$

where the coefficients A_{nm} are fixed by the two-dimensional Fourier transform of the initial profile

$$f(x, y) = \sum_n \sum_m A_{nm} \sin\left(n\frac{\pi x}{L_x}\right) \sin\left(m\frac{\pi y}{L_y}\right)$$

and determined as area integrals

$$A_{nm} = \frac{2}{L_x}\frac{2}{L_y} \int_0^{L_x} dx \int_0^{L_y} dy f(x, y) \sin\left(n\frac{\pi x}{L_x}\right) \sin\left(m\frac{\pi y}{L_y}\right).$$

Let us consider as a rectangular domain is $[0, \pi] \times [0, \pi]$, i.e. $L_x = L_y = \pi$ and a constant initial profile $f(x, y) = 1$. Then, the field at time t starting from this constant profile and homogeneous boundary conditions is given as

$$u(x, y, t) = \sum_n \sum_m A_{nm} e^{-(n^2+m^2)t} \sin(nx) \sin(my)$$

where

$$A_{nm} = \frac{4}{\pi^2} \int_0^\pi dx \sin(nx) \int_0^\pi dy \sin(my) \tag{6}$$

$$= \frac{4^2}{\pi^2} \frac{1}{nm} \tag{7}$$

Hence, the solution reads as

$$u(x, y, t) = \frac{4^2}{\pi^2} \sum_n \sum_m \frac{1}{mn} e^{-(n^2+m^2)t} \sin(nx) \sin(my).$$

This formula can be generalized to 3D as

$$u(x, y, z, t) = \frac{4^3}{\pi^3} \sum_n \sum_m \sum_q \frac{1}{mnq} e^{-(n^2+m^2+q^2)t} \sin(nx) \sin(my) \sin(qz).$$

2 Lecture 25: Separation of Variable Method Non-Cartesian Coordinates

The method of separation of variables may be used to solve problems in higher dimensions. Spatial symmetries are important when choosing appropriate spatial coordinates. For example, problems defined on domains with angular symmetry (where the system is invariant under rotations) tend to have solutions that reflect these same symmetries. Recognizing such symmetries can guide us toward finding solutions more efficiently and simplifying complex problems.

In this lecture, we will apply the separation of variable method to solve pde's on a disk or on a sphere. As we do so, we will encounter some familiar special ode's, such as the Euler-Cauchy equation, Legendre equation, and Bessel equation.

2.1 Laplace Equation: Spherical Coordinates

Let us consider a scalar function of three independent variables $u = u(r, \phi, \theta)$ corresponding to the spherical coordinates to represent a point in a three-dimensional space, i.e. radial distance r to origin, inclination angle $\theta \in [0, \pi]$ from the z-axis and azimuth angle $\phi \in [0, 2\pi]$.

A pde satisfied by this function contains spatial derivatives which in the Cartesian (x, y, z) coordinates are

$$\frac{\partial u}{\partial x}, \quad \frac{\partial u}{\partial y}, \quad \frac{\partial u}{\partial z}$$

and of higher order.

To find the differential equation in spherical coordinates, we make use of the transformation of variables

$$x = r \cos \phi \sin \theta,$$
$$y = r \sin \phi \sin \theta,$$
$$z = r \cos \theta.$$

Let us denote the unit vectors **i**, **j**, **k** along the coordinate axes in the (x, y, z)-space transform, such that the corresponding unit vectors for the spherical coordinate system are

$$\hat{\mathbf{r}} = \mathbf{i}\sin\theta\cos\phi + \mathbf{j}\sin\theta\sin\phi + \mathbf{k}\cos\theta \tag{8}$$

$$\hat{\boldsymbol{\theta}} = \mathbf{i}\cos\theta\cos\phi + \mathbf{j}\cos\theta\sin\phi - \mathbf{k}\sin\theta \tag{9}$$

$$\hat{\boldsymbol{\phi}} = -\mathbf{i}\sin\phi + \mathbf{j}\cos\phi. \tag{10}$$

A point in space can be uniquely identified as the intersection of three mutually perpendicular planes. The normal vectors to these planes relate to an orthogonal coordinate system. For instance, we can locate a point in the (x, y, z) coordinates by the intersection between the plane $x = constant$, the plane $y = constant$ and the plane of $z = constant$. The same point can also be located in spherical coordinates by the interactions of the three planes $r = constant$, $\theta = constant$, $\phi = constant$.

Using these coordinate transformations, one can show that the gradient vector

$$\nabla u = \frac{\partial u}{\partial x}\mathbf{i} + \frac{\partial u}{\partial y}\mathbf{j} + \frac{\partial u}{\partial z}\mathbf{k}$$

can be expressed in spherical coordinates as

$$\nabla u = \frac{\partial u}{\partial r}\hat{\mathbf{r}} + \frac{1}{r}\frac{\partial u}{\partial \theta}\hat{\boldsymbol{\theta}} + \frac{1}{r\sin\theta}\frac{\partial u}{\partial \phi}\hat{\boldsymbol{\phi}}.$$

The Laplacian operator defined as the divergence of the gradient field $\nabla \cdot \nabla$ is expressed in Cartesian coordinates

$$\nabla^2 u = \frac{\partial^2 u}{\partial x^2} + \frac{\partial^2 u}{\partial y^2} + \frac{\partial^2 u}{\partial z^2}.$$

In spherical coordinates, the Laplacian of a function $u(r, \phi, \theta)$ takes the form

$$\nabla^2 u = \frac{1}{r^2}\frac{\partial}{\partial r}\left(r^2\frac{\partial u}{\partial r}\right) + \frac{1}{r^2}\frac{1}{\sin\theta}\frac{\partial}{\partial \theta}\left(\sin\theta\frac{\partial u}{\partial \theta}\right) + \frac{1}{r^2}\frac{1}{\sin^2\theta}\frac{\partial^2 u}{\partial \phi^2}$$

By setting the Laplacian of u equal to zero, we obtain the *Laplace equation*

$$\nabla^2 u = 0.$$

The Laplace equation is often associated with stationary or steady-state phenomena.

Azimuthal Symmetry: We seek to find a solution $u = u(r, \theta)$, that means u is independent of the azimuth angle ϕ about the z-axis, thus

$$\frac{\partial u}{\partial \phi} = 0.$$

Then, the 3D Laplace equation simplifies to

$$\frac{\partial}{\partial r}\left(r^2 \frac{\partial u}{\partial r}\right) + \frac{1}{\sin\theta} \frac{\partial}{\partial \theta}\left(\sin\theta \frac{\partial u}{\partial \theta}\right) = 0. \tag{11}$$

In addition, we use radial boundary conditions given by

$$u(r=R, \theta) = f(\theta), \qquad \lim_{r\to\infty} u(r, \theta) = 0$$

meaning that on the surface of a sphere of radius R, the function u equals a given profile $f(\theta)$ that may depend on the angle θ, and as the radius becomes infinite, the function decays to zero. The field $u(r, \theta)$ may represent, for instance, the electrostatic potential induced by a charged spherical object that is empty inside.

By the separation of variables, we construct an ansatz solution as a product of two functions

$$u(r, \theta) = G(r) \cdot H(\theta)$$

and insert this into Eq. (11) to obtain the corresponding relation satisfied by them

$$\frac{1}{G}\frac{d}{dr}\left(r^2 \frac{dG}{dr}\right) = -\frac{1}{H\sin\theta}\frac{d}{d\theta}\left(\sin\theta \frac{dH}{d\theta}\right) \equiv k$$

where k is the proportionality constant determined by the boundary conditions.

The second order ode in r,

$$\frac{d}{dr}\left(r^2 \frac{dG}{dr}\right) = kG \rightarrow r^2 G'' + 2r G' - kG = 0,$$

is the **Euler-Cauchy equation**. Without loss of generality, we take $k = n(n+1)$ with $n = 0, 1, 2, \ldots$ such that the above equation becomes

$$r^2 G'' + 2r G' - n(n+1) G = 0.$$

We seek for independent solutions of the form r^λ with λ determined by the characteristic equation:

$$\lambda(\lambda - 1) + 2\lambda - n(n+1) = 0 \Rightarrow \lambda_1 = n, \quad \lambda_2 = -(n+1).$$

Hence, the two independent solutions are parameterised by n and given by:

$$G_n(r) = r^n, \qquad \tilde{G}_n(r) = r^{-(n+1)}.$$

The ode in θ is

$$\frac{1}{\sin\theta}\frac{d}{d\theta}\left(\sin\theta\frac{dH}{d\theta}\right) + n(n+1)H = 0. \tag{12}$$

We use the change of variable

$$\omega = \cos\theta \Rightarrow \sin^2\theta = 1 - \omega^2,$$

such that the derivative with respect to θ transforms as

$$\frac{d}{d\theta} = \frac{d\omega}{d\theta}\frac{d}{d\omega} = -\sin\theta\frac{d}{d\omega}.$$

Equation (12) can be rewritten in ω as

$$-\frac{d}{d\omega}\left(-\sin\theta^2\frac{dH}{d\omega}\right) + n(n+1)H = 0 \Rightarrow$$
$$\frac{d}{d\omega}\left((1-\omega^2)\frac{dH}{d\omega}\right) + n(n+1)H = 0,$$

which is the **Legendre equation**

$$(1-\omega^2)H'' - 2\omega H' + n(n+1)H = 0,$$

defined in the interval $\omega \in [-1, 1]$. We have solved this equation using the power series expansion method. The solution which is finite and equal to 1 at the boundary points $\omega = \pm 1$ is given by the Legendre polynomial of order n,

$$H(\omega) = P_n(\omega) \to H(\theta) = P_n(\cos\theta).$$

In our case, $n = 0, 1, \ldots$ and labels the infinite number of independent solutions of the Laplace equation are given by

$$u_n(r,\theta) = r^n P_n(\cos\theta), \qquad \tilde{u}_n(r,\theta) = r^{-(n+1)} P_n(\cos\theta).$$

Thus, the general solution can be written as a series expansion in terms of these eigenfunctions, namely

$$u(r,\theta) = \sum_{n=0}^{\infty}\left[A_n r^n + B_n r^{-(n+1)}\right] P_n(\cos\theta), \tag{13}$$

with constants A_n and B_n determined by the boundary conditions and the orthogonality relation of the Legendre polynomials. Recall that

$$\int_{-1}^{1} d\omega\, P_n(\omega) P_m(\omega) = \frac{2}{2n+1} \delta_{n,m}$$

or equivalently using the θ variable

$$\int_{0}^{\pi} d\theta\, \sin\theta\, P_n(\cos\theta) P_m(\cos\theta) = \frac{2}{2n+1} \delta_{n,m}$$

Application to Potential Fields: Let us apply this solution to find the electrostatic potential $u(r,\theta)$ inside and outside of a surface of radius R for a known profile of the electrostatic potential on the surface namely

$$u(R,\theta) = f(\theta).$$

Outside Solution: The boundary condition

$$\lim_{r \to \infty} u(r,\theta) = 0$$

implies that the field is zero at infinity, hence

$$A_n = 0.$$

Thus, the solution from Eq. (13) simplifies to

$$u(r,\theta) = \sum_{n=0}^{\infty} B_n r^{-(n+1)} P_n(\cos\theta), \qquad r \geq R$$

The coefficients B_n are determined from the orthogonality condition applied to the radial boundary condition at $r = R$,

$$u(R,\theta) = f(\theta).$$

Namely,

$$\int_{0}^{\pi} d\theta\, \sin\theta\, u(R,\theta) P_m(\cos\theta) = \sum_{n=0}^{\infty} B_n R^{-(n+1)} \int_{0}^{\pi} d\theta\, \sin\theta\, P_n(\cos\theta) P_m(\cos\theta)$$

$$= B_m R^{-(m+1)} \frac{2}{2m+1}$$

Hence, the coefficients are computed as

$$B_n = R^{(n+1)}\frac{(2n+1)}{2}\int_0^\pi d\theta \sin(\theta) f(\theta) P_n(\cos\theta).$$

Inside Solution: For a finite value at the origin, the coefficients of diverging terms $r^{-(n+1)}$ must vanish, hence

$$B_n = 0, \quad r \leq R,$$

such that the inside solution is expanded as

$$u(r,\theta) = \sum_{n=0}^\infty A_n r^n P_n(\cos\theta), \quad r \leq R$$

The coefficients A_n are determined by the profile on the surface of radius R

$$\int_0^\pi d\theta \sin\theta\, u(R,\theta) P_m(\cos\theta) = \sum_{n=0}^\infty A_n R^n \int_0^\pi d\theta \sin\theta\, P_n(\cos\theta) P_m(\cos\theta)$$

$$= A_m R^m \frac{2}{2m+1}$$

Hence,

$$A_n = R^{-n}\frac{(2n+1)}{2}\int_0^\pi d\theta \sin(\theta) f(\theta) P_n(\cos\theta)$$

For a constant field on the surface $f(\theta) = 1$, the integral that determines the coefficients A_n and B_n reduces to the integral over the Legendre polynomials,

$$\int_{-1}^1 d\omega\, P_n(\omega) = 2\delta_{n,0}.$$

All the odd polynomials integrate to zero, while for the even polynomials only $P_0(\omega) = 1$ leads to a nonzero integral. The solution of the electrostatic potential for this constant distribution on the spherical surface is

$$u(r,\theta) = \begin{cases} \frac{R}{r}, & r \geq R \\ 1, & r \leq R. \end{cases}$$

2.2 Diffusion Equation in Polar Coordinates

For the disk geometry, we may choose the polar coordinates

$$x = r\cos\theta, \quad y = r\sin\theta.$$

The 2D Laplace operator in polar coordinates reads as

$$\nabla^2 u = \frac{1}{r}\frac{\partial}{\partial r}\left(r\frac{\partial u}{\partial r}\right) + \frac{1}{r^2}\frac{\partial^2 u}{\partial \theta^2}$$

where $u = u(r, \theta)$. We are seeking for isotropic solution, namely solutions that are independent on the angle θ

$$\frac{\partial u}{\partial \theta} = 0.$$

Let us consider the diffusion of a field $u(r, t)$ within a disk,

$$\frac{\partial u}{\partial t} = c^2 \frac{1}{r}\frac{\partial}{\partial r}\left(r\frac{\partial u}{\partial r}\right).$$

We assume *homogeneous* boundary condition on the circumference of the disk

$$u(r = a, t) = 0, \text{ on the boundary of the disk, for any } t \geq 0 \quad (14)$$

and for an initial radial profile

$$u(r, 0) = f(r).$$

We use the ansatz solution
$$u(t, r) = F(r)G(t)$$

into the diffusion equation, and arrive at

$$\frac{1}{F}\frac{1}{r}\frac{d}{dr}\left(r\frac{dF}{dr}\right) = \frac{1}{c^2}\frac{1}{G}\frac{dG}{dt} = -k^2.$$

The time evolution reduces to a first order ode

$$\frac{\dot{G}(t)}{G(t)} = -k^2 c^2 t,$$

with the independent solution given by the exponential decay

$$G_1(t) = e^{-k^2 c^2 t}.$$

The radial evolution corresponds to

$$\frac{d}{dr}\left(r\frac{dF}{dr}\right) - k^2 r F = 0,$$

which is a specific form of the **Bessel** equation

$$x\frac{d}{dx}\left(x\frac{dy}{dr}\right) + (x^2 - n^2)y = 0$$

for $n = 0$ and $x = kr$. We have solved this equation using the Frobenius method and found that one of the independent solutions is given by the *Bessel* function of the first kind of order n $J_n(kr)$. In our case, $n = 0$. It turns out that the other independent solution is singular at the origin, thus does not contribute to the general solution expansion, i.e. the eigenfunctions depend only on

$$J_0(kr) = \sum_{n=0}^{\infty}(-1)^n \frac{1}{(n!)^2}\left(\frac{kr}{2}\right)^{2n}.$$

The homogeneous condition on the boundary of the disk at $r = a$, implies that the Bessel function J_0 must vanish $J_0(ka) = 0$. However, we notice that J_0 has infinitely many zeros at positions μ_n where $n = 0, 1, 2, \ldots$. This determines the allowed values of the proportionality constant k from the position of the zeros, namely that

$$k = \frac{\mu_n}{a}, \quad n \in \mathcal{N}.$$

Thus, the solution is a series expansion

$$u(r,t) = \sum_{n=0}^{\infty} A_n J_0\left(\mu_n \frac{r}{a}\right) e^{-\gamma_n t}$$

where $\gamma_n = \frac{\mu_n^2 c^2}{a^2}$ is the decay rate.

Applying this expression to the initial profile $u(r, t) = f(r)$, we find that the $f(r)$ is represented as an infinite series expansion

$$f(r) = \sum_{n=0}^{\infty} A_n J_0\left(\mu_n \frac{r}{a}\right).$$

The Bessel function of order zero J_0 satisfies this orthogonality relation

$$\int_0^a dr\, r\, J_0\left(\mu_n \frac{r}{a}\right) J_0\left(\mu_m \frac{r}{a}\right) = \frac{a^2}{2} J_1^2(\mu_n)\, \delta_{n,m}$$

where $J_1(\mu_n)$ is the first order Bessel function evaluated at the position of the n-th zero μ_n of J_0. J_1 has its own set of zeros that are different from those of J_0, thus $J_1(\mu_n) \neq 0$. We use this property to determine the coefficients A_n as

$$\int_0^a dr\, r J_0\left(\mu_m \frac{r}{a}\right) f(r) = \sum_n A_n \int_0^a dr\, r J_0\left(\mu_m \frac{r}{a}\right) J_0\left(\mu_n \frac{r}{a}\right)$$

$$= \frac{a^2}{2} A_m J_1^2(\mu_m)$$

which implies that

$$A_m = \frac{2}{a^2} \frac{1}{J_1^2(\mu_m)} \int_0^a dr\, r J_0\left(\mu_m \frac{r}{a}\right) f(r).$$

For a uniform initial profile $f(r) = 1$ on a disk of radius $a = 1$, the corresponding coefficients are

$$A_m = 2\frac{1}{J_1^2(\mu_m)} \int_0^1 dr\, r J_0(\mu_m r)$$

$$= \frac{2}{\mu_m} \frac{1}{J_1^2(\mu_m)} J_1(\mu_m)$$

$$= \frac{2}{\mu_m} J_1(\mu_m)$$

and the solution at radius r and t is then

$$u(r,t) = \sum_{m=0}^{\infty} \frac{2}{\mu_m} \frac{J_0(\mu_m r)}{J_1(\mu_m)} e^{-\mu_m^2 t},$$

where $c = 1$. In practice, this could represent the temperature field (in some rescaled units) inside the disk starting from a uniform value and fixing the temperature to zero on the circumference of the disk. What happens is that the heat is diffused out of the system and as result the temperature inside the disk will gradually lower with time.

3 Lecture 26: Integral Transform Method

While the separation of variables method is a valuable tool, it is often not suitable when forcing term is non-zero under non-homogeneous boundary conditions. In such cases, we may rely on the integral transform method. Techniques like the Fourier or Laplace transforms are particularly useful for tackling linear boundary value or initial value problems.

In this lecture, we will focus on the integral transform approach and the Green's function method.

3.1 Laplace Transform Method

The Laplace transform method is tailored to linear, initial value problems.

Transport Equation: Let us consider a homogeneous first order PDE given by

$$x\frac{\partial u}{\partial t} + \frac{\partial u}{\partial x} = 0$$

with the **homogeneous initial condition**

$$u(x, 0) = 0$$

and the non-homogeneous boundary condition at $x = 0$

$$u(0, t) = t, \quad t \geq 0$$

Let us denote the Laplace transform of the field $u(x, t)$ as

$$\mathcal{L}[u(x, t)] = U(x, s).$$

Using the homogeneous initial condition, it follows that the L.T. of the time derivative of $u(x, t)$ is then

$$\mathcal{L}[\partial_t u(x, t)] = sU(x, s).$$

The first step in solving this PDE is to Laplace transform the equation with respect to time t, which leads to this differential equation

$$sxU + \frac{\partial U}{\partial x} = 0$$

Since there are no derivatives with respect to s, this is a first order ode in x,

$$\frac{dU}{U} = -sx\,dx,$$

which leads to the general solution

$$U(x, s) = c(s)e^{-sx^2/2}.$$

3 Lecture 26: Integral Transform Method

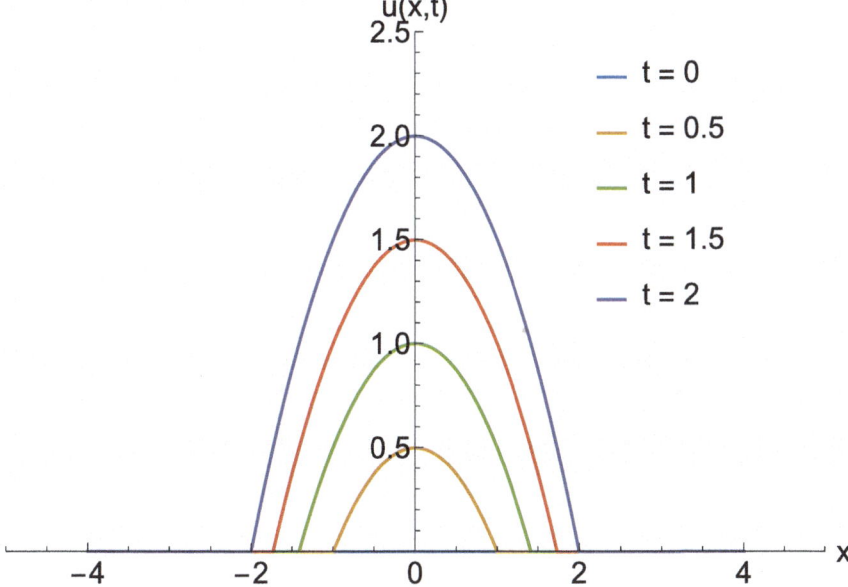

Fig. 3 The profile of u(x,t) as function of x for different values of t

The integration factor $c(s)$ that may depend on s and is fixed by the boundary condition at $x = 0$ where the Laplace transform function $U(0, s) = c(s)$ must equal the Laplace transform of t, namely

$$c(s) = \mathcal{L}[t] = \frac{1}{s^2}.$$

Hence, the solution in the s-space is

$$U(x, s) = \frac{1}{s^2} e^{-sx^2/2}$$

and its inverse Laplace transform gives us the specific solution to this problem,

$$u(x, t) = \mathcal{L}^{-1}\left[\frac{1}{s^2} e^{-sx^2/2}\right].$$

Using the t-shifting theorem,

$$\mathcal{L}^{-1}\left[F(s) e^{-sa}\right] = f(t - a) H(t - a),$$

with $a = x^2/2$ and $F(s) = \frac{1}{s^2} = \mathcal{L}[t]$, we then obtain the final solution given as (see Fig. 3)

$$u(x,t) = \left(t - \frac{x^2}{2}\right) H\left(t - \frac{x^2}{2}\right)$$

where $H(x)$ is the Heaviside function. Using the definition of the Heaviside function, we can write the solution equivalently as

$$u(x,t) = \begin{cases} 0, & t < \frac{x^2}{2} \\ t - \frac{x^2}{2}, & t \geq \frac{x^2}{2} \end{cases} \qquad (15)$$

We may notice that the final solution is not consistent with the ansatz solution from the separation of variable method, and seeking to solve the problem using that ansatz would only give us the trivial solution (zero).

1D Wave Equation: Let us consider again the 1D wave equation

$$\frac{\partial^2 u}{\partial t^2} = c^2 \frac{\partial^2 u}{\partial x^2}$$

corresponding to travelling waves with the speed c and along an infinite string. We take the *homogeneous initial conditions*

$$u(x,0) = 0, \quad \partial_t u(x,0) = 0$$

and the boundary conditions on the string which extends from $x = 0$ to $x \to \infty$, given by:

$$u(0,t) = \sin(t) H(2\pi - t) = \begin{cases} \sin(t), & 0 \leq t \leq 2\pi \\ 0, & \text{otherwise} \end{cases}, \quad \lim_{x \to \infty} u(x,t) = 0, \quad t \geq 0.$$

These boundary conditions mean that the end of the string at $x = 0$ oscillates in time for one period and then it is fixed at 0 for the remaining time. The other end at infinity is fixed at all times. In other words, we set up a wave at one end and solve for the traveling wave across the string.

Again, we denote the Laplace transform of the deformation field u with respect to time t as

$$U(x,s) = \mathcal{L}[u(x,t)]$$

The homogeneous initial conditions simplify the expressions of the L.T. of the second derivative in time, such that the L.T. of the wave equation reads as

$$\frac{d^2 U}{dx^2} - \frac{s^2}{c^2} U = 0.$$

This is a second order ode in x with the solution

$$U(x, s) = A(s)e^{sx/c} + B(s)e^{-sx/c},$$

where the integration coefficients $A(s)$ and $B(s)$ may depend on s and are fixed by the boundary conditions at x. Namely, using the boundary condition at infinity of zero displacement,

$$\lim_{x \to \infty} U(x, s) = 0 \to A(s) = 0,$$

we find that

$$U(x, s) = B(s)e^{-sx/c}.$$

Evaluating this solution at $x = 0$, we find that $U(0, s) = B(s)$ which must equal the Laplace transform of the boundary condition profile at $x = 0$, hence

$$B(s) = \mathcal{L}[u(0, t)]$$

By the t-shifting theorem, the inverse Laplace transform gives us the specific solution as

$$u(x, t) = u\left(0, t - \frac{x}{c}\right) H\left(t - \frac{x}{c}\right),$$

Using the initial profile, we can rewrite the solution as

$$u(x, t) = \sin\left(t - \frac{x}{c}\right) H\left(2\pi + \frac{x}{c} - t\right) H\left(t - \frac{x}{c}\right),$$

which is equivalent to

$$u(x, t) = \begin{cases} \sin\left(t - \frac{x}{c}\right), & \frac{x}{c} \leq t \leq 2\pi + \frac{x}{c} \\ 0, & \text{otherwise} \end{cases}$$

3.2 Fourier Transform Method

We now exemplify how to apply the Fourier transform method for boundary value problems, where the function decays sufficiently fast at $\pm \infty$. Let us consider the diffusion equation in one spatial dimension,

$$\frac{\partial u}{\partial t} = c^2 \frac{\partial^2 u}{\partial x^2}$$

with a non-homogeneous initial condition

$$u(x, 0) = b\delta(x)$$

and **homogeneous boundary conditions at $\pm \infty$,**

$$\lim_{x \to \pm\infty} u(x,t) = 0, \qquad t \geq 0.$$

Let us denote the Fourier transform with respect to x of the function itself by

$$\mathcal{F}[u(x,t)] = \hat{u}(k,t).$$

We apply the Fourier transform of the diffusion equation and arrive at a first order ode in time,

$$\frac{d\hat{u}}{dt} + c^2 k^2 \hat{u} = 0,$$

which has the solution given by a decaying exponential

$$\hat{u}(k,t) = A(k)e^{-c^2 k^2 t},$$

with the integration coefficient $A(k)$ that may depend on k through the initial condition. Namely, setting $t = 0$ in the above equation and equating it with the Fourier transform of the initial profile, we have

$$A(k) = \mathcal{F}[b\delta(x)] = \frac{b}{2\pi},$$

thus the solution is a Gaussian function of k in the Fourier space,

$$\hat{u}(k,t) = \frac{b}{2\pi} e^{-c^2 k^2 t}.$$

By applying the inverse Fourier transform on $\hat{u}(k,t)$, we then find the solution in the x variable given by

$$u(x,t) = \frac{b}{2\pi} \mathcal{F}^{-1}\left[e^{-c^2 k^2 t}\right]$$

$$= \frac{b}{2\pi} \int_{-\infty}^{\infty} dk\, e^{-c^2 k^2 t} e^{ikx}.$$

By completing the square in the exponent and using the Gaussian integral formula, we arrive at

$$u(x,t) = \frac{b}{2\pi} e^{-x^2/(4c^2 t)} \int_{-\infty}^{\infty} dk\, e^{-(c\sqrt{t}k + ix/\sqrt{4c^2 t})^2}$$

$$= \frac{b}{\sqrt{4\pi c^2 t}} e^{-x^2/(4c^2 t)},$$

3 Lecture 26: Integral Transform Method

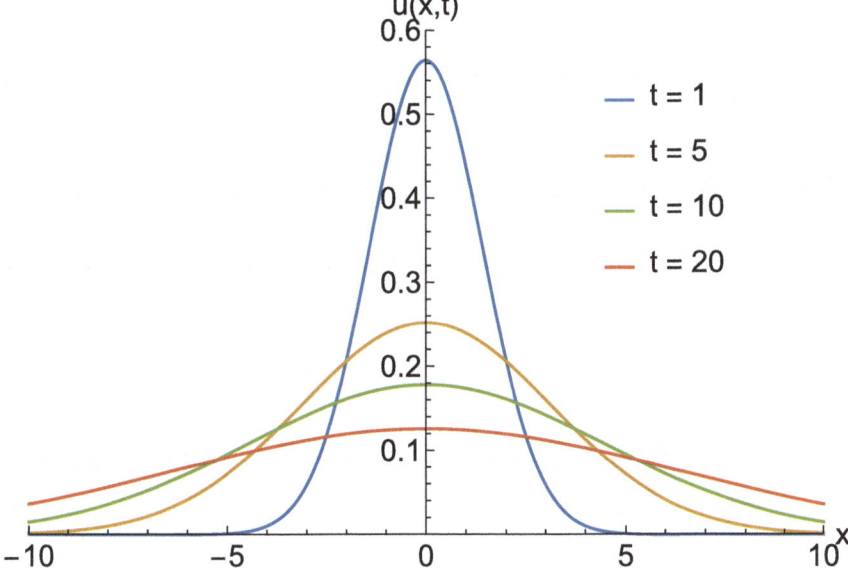

Fig. 4 The Gaussian profile at different times for $c = 1$ and $b = 2$

which is a Gaussian peak centered at origin and with a time-dependent width that increases with time (as shown in Fig. 4). This represents the diffusion process which spreads out the field in space and time. Notice that the Gaussian distribution is normalized at all times,

$$\int_{-\infty}^{+\infty} dx\, u(x,t) = 1, \tag{16}$$

which follows straightforwardly using the Gaussian integral

$$\int_{-\infty}^{+\infty} dx\, e^{-\alpha x^2} = \sqrt{\frac{\pi}{\alpha}}.$$

The Fourier transform is a powerful method. In higher dimensions, we perform a Fourier transform with respect to each spatial component. For example, the position vector $\vec{r}$ with coordinates (x, y, z) in the 3D space has a corresponding conjugate wave vector $\vec{k}$ with wavenumber coordinates (k_x, k_y, k_z) in the Fourier space. Thus, the Fourier transform of a field $u(\vec{r}, t)$ is another field $\hat{u}(\vec{k}, t)$ in the $\vec{k}$-space and given as

$$\hat{u}(\vec{k},t) = \frac{1}{(2\pi)^3} \int_{-\infty}^{+\infty} dx \int_{-\infty}^{+\infty} dy \int_{-\infty}^{+\infty} dz u(x,y,z,t) e^{-ik_x x} e^{-ik_y y} e^{-ik_z z}$$

$$= \frac{1}{(2\pi)^3} \iiint d^3\vec{r} u(\vec{r},t) e^{-i\vec{k}\cdot\vec{r}}.$$

When the problem has a *spherical symmetry* whereby the field depends only on the radius $r = |\vec{r}|$, i.e. $u(\vec{r},t) = u(r,t)$, this symmetry carries over to the Fourier space, i.e. $\hat{u}(\vec{k},t) = \hat{u}(k,t)$. We can make use of the transformation to spherical variables to obtain

$$\hat{u}(k,t) = \frac{1}{(2\pi)^2} \int_0^{+\infty} dr r^2 u(r,t) \int_0^{\pi} d\theta \sin\theta e^{-ikr\cos\theta} \tag{17}$$

$$= \frac{1}{(2\pi)^2} \int_0^{+\infty} dr r^2 u(r,t) \int_{-1}^{1} dt e^{-ikrt} \tag{18}$$

$$= \frac{1}{(2\pi)^2} \int_0^{+\infty} dr r^2 u(r,t) \frac{1}{ikr} \left(e^{ikr} - e^{-ikr} \right) \tag{19}$$

$$= \frac{1}{2\pi^2 k} \int_0^{+\infty} dr r u(r,t) \sin(kr). \tag{20}$$

3.3 Green's Function Method

The Green's function method applies to non-homogeneous, linear differential equations. As an example, we apply it to solve the Poisson equation for $u(x,y,z)$, namely

$$\frac{\partial^2 u(x,y,z)}{\partial x^2} + \frac{\partial^2 u(x,y,z)}{\partial y^2} + \frac{\partial^2 u(x,y,z)}{\partial z^2} = f(x,y,z),$$

with the homogeneous boundary conditions

$$\lim_{x \to \pm\infty} u(x,y,z) = 0, \forall y,z,$$

$$\lim_{y \to \pm\infty} u(x,y,z) = 0, \forall x,z,$$

$$\lim_{z \to \pm\infty} u(x,y,z) = 0, \forall x,y.$$

3 Lecture 26: Integral Transform Method

The corresponding equation satisfied by the Green's function $G(x, y, z, x', y', z')$ is obtained by replacing the forcing term with a product of Dirac delta functions for each variable. Let us introduce the Dirac delta function

$$\delta(x - x') \cdot \delta(y - y') \cdot \delta(z - z') \equiv \delta^3(\vec{r} - \vec{r}')$$

such that,

$$\nabla^2 G(x, y, z, x', y', z') = \delta^3(\vec{r} - \vec{r}').$$

By the convolution theorem, the solution of the Poisson equation with a source $f(x, y, z)$ follows as

$$u(x, y, z) = \int_{-\infty}^{\infty} dx' \int_{-\infty}^{\infty} dy' \int_{-\infty}^{\infty} dz' G(x, y, z, x', y', z') f(x', y', z').$$

We can solve the equation satisfied by the Green's function using the Fourier transform method. We denote the 3D Fourier transform of the Green function as

$$\hat{G}(k_x, k_y, k_z, x', y', z') = \mathcal{F}[G]$$
$$= \frac{1}{(2\pi)^3} \int_{-\infty}^{\infty} dx \int_{-\infty}^{\infty} dy \int_{-\infty}^{\infty} dz G(x, y, z, x', y', z') e^{-i(k_x x + k_y y + k_z z)}.$$

We also need to Fourier transform the product of the delta functions

$$\mathcal{F}[\delta^3(\vec{r} - \vec{r}')] = \frac{1}{(2\pi)^3} \iiint d^3 r \, \delta^3(\vec{r} - \vec{r}') e^{-i\vec{k}\cdot\vec{r}}$$
$$= \frac{1}{(2\pi)^3} e^{-i\vec{k}\cdot\vec{r}'}.$$

The Fourier transforms of the spatial derivatives are obtained by successive integration by parts and given as

$$\mathcal{F}[\partial_x^2 G] = -k_x^2 \hat{G}, \quad \mathcal{F}[\partial_y^2 G] = -k_y^2, \quad \mathcal{F}[\partial_z^2 G] = -k_z^2$$

Putting it all together, we find that the Poisson equation for the Green's function in the Fourier space reduces to an algebraic equation

$$-(k_x^2 + k_y^2 + k_z^2)\hat{G} = \frac{1}{(2\pi)^3} e^{-i(k_x x' + k_y y' + k_z z')}$$

which determines $\hat{G}$ as

$$\hat{G} = -\frac{1}{(2\pi)^3} \frac{1}{k^2} e^{-i\vec{k}\cdot\vec{r}'}$$

By applying the inverse F.T. on $\hat{G}$ and writing the triple integral over $\vec{k}$ in spherical coordinates, we find an expression for the Green function in the $\vec{r}$-space given as

$$G(\vec{r},\vec{r}') = \iiint d^3k \hat{G}(\vec{k},\vec{r}') e^{i\vec{k}\cdot\vec{r}}$$

$$= -\frac{1}{(2\pi)^3} \int_0^{2\pi} d\phi \int_0^{\pi} d\theta \sin\theta \int_0^{\infty} dk k^2 \frac{1}{k^2} e^{i\vec{k}\cdot(\vec{r}-\vec{r}')}$$

$$= -\frac{2\pi}{(2\pi)^3} \int_0^{\pi} d\theta \sin\theta \int_0^{\infty} dk k^2 \frac{1}{k^2} e^{ik|\vec{r}-\vec{r}'|\cos\theta},$$

where we have integrated out the azimuthal angle ϕ and aligned the $\vec{k}$-coordinate system so that the vector $\vec{r} - \vec{r}'$ is along the k_z direction such that the inner product of $\vec{k}$ with $\vec{r} - \vec{r}'$ is determined by the inclination angle θ. Let us now denote by $a = |\vec{r} - \vec{r}'| > 0$ the magnitude of this vector. We can perform the integral over the inclination angle θ, by using the substitution $t = \cos\theta$, namely

$$G(a) = -\frac{1}{(2\pi)^2} \int_0^{\infty} dk \int_{-1}^{1} dt\, e^{ikat}$$

$$= -\frac{1}{(2\pi)^2} \int_0^{\infty} dk \frac{1}{ika} \left[e^{ika} - e^{-ika}\right]$$

$$= -\frac{1}{(2\pi)^2} \frac{2}{a} \int_0^{\infty} dk \frac{\sin(ka)}{k}.$$

Since the integrand is even, we rewrite the integral as

$$G(a) = -\frac{1}{(2\pi)^2} \frac{1}{a} \int_{-\infty}^{\infty} dk \frac{\sin(ka)}{k}$$

$$= -\frac{1}{(2\pi)^2} \frac{1}{a} \int_{-\infty}^{\infty} dk \frac{1}{2ik} \left[e^{ika} - e^{-ika}\right].$$

This integrand has a singular behavior at $k = 0$, but we can evaluate its principal value. We can deal with it by extending the integral into the complex plane in the complex variable w such that $Re(w) = k$ and apply the principal value theorem. Thus,

3 Lecture 26: Integral Transform Method

$$\int_{-\infty}^{\infty} dk \frac{\sin(ka)}{k} = \frac{1}{2i} \text{P.V.} \int_{-\infty}^{\infty} dk \frac{e^{ika}}{k} - \frac{1}{2i} \text{P.V.} \int_{-\infty}^{\infty} dk \frac{e^{-ika}}{k}$$

$$= \frac{\pi}{2} + \frac{\pi}{2} = \pi.$$

Hence, the Green's function of the Poisson equation takes this simple form

$$G(\vec{r}, \vec{r}') = -\frac{1}{4\pi} \frac{1}{|\vec{r} - \vec{r}'|}.$$

It is spherical symmetric, thus depends only on the distance between two points $G(|\vec{r} - \vec{r}'|)$. The integral solution to the Poisson equation with a given source $f(\vec{r})$ is then the convolution integral

$$u(\vec{r}) = \iiint d^3\vec{r}' G(|\vec{r} - \vec{r}'|) f(\vec{r}').$$

This is the typical integral solution for the gravitational field or electrostatic field induced by a given source in 3D space.

Correction to: Analytical Methods in Physics

Correction to:
Chapters 3, 4 and 5 in: L. Angheluta, *Analytical Methods in Physics*, **https://doi.org/10.1007/978-3-031-77053-1**

In the original version of the book, the following belated corrections have been updated in chapters 3, 4, and 5: The corrections in Examples 13 and 14 have been incorporated. The book and the correction chapters has been updated with the changes.

The updated versions of these chapters can be found at
https://doi.org/10.1007/978-3-031-77053-1_3
https://doi.org/10.1007/978-3-031-77053-1_4
https://doi.org/10.1007/978-3-031-77053-1_5

© The Author(s), under exclusive license to Springer Nature Switzerland AG 2025
L. Angheluta, *Analytical Methods in Physics*,
https://doi.org/10.1007/978-3-031-77053-1_7

References

1. George B Arfken, Hans J Weber, and Frank E Harris. *Mathematical methods for physicists: a comprehensive guide*. Academic press, 2011.
2. Mary L Boas and Philip Peters. Mathematical methods in the physical sciences. *American Journal of Physics*, 52(6):572–573, 1984.
3. William E Boyce, Richard C DiPrima, and Douglas B Meade. *Elementary differential equations*. John Wiley & Sons, 2017.
4. Cyrus D Cantrell. *Modern mathematical methods for physicists and engineers*. Cambridge University Press, 2000.
5. Philippe Dennery and Andrzej Krzywicki. *Mathematics for physicists*. Courier Corporation, 1996.
6. Gerald B Folland. *Fourier analysis and its applications*, volume 4. American Mathematical Soc., 2009.
7. Sadri Hassani. *Mathematical physics: a modern introduction to its foundations*. Springer Science & Business Media, 2013.
8. John Mathews and Russell Howell. *Complex analysis for mathematics and engineering*. Jones & Bartlett Publishers, 2012.
9. Vasilis Pagonis and Christopher Wayne Kulp. *Mathematical Methods using Python: Applications in Physics and Engineering*. CRC Press, 2024.
10. Kenneth Franklin Riley, Michael Paul Hobson, and Stephen John Bence. *Mathematical methods for physics and engineering: a comprehensive guide*. Cambridge university press, 2006.
11. Roel Snieder and Kasper Van Wijk. *A guided tour of mathematical methods for the physical sciences*. Cambridge University Press, 2015.
12. Ian Stewart and David Tall. *The foundations of mathematics*. OUP Oxford, 2015.
13. Michael Stone and Paul Goldbart. *Mathematics for physics: a guided tour for graduate students*. Cambridge University Press, 2009.
14. Kwong-Tin Tang. *Mathematical methods for engineers and scientists*, volume 2. Springer, 2006.
15. Wolfgang Yourgrau and Stanley Mandelstam. *Variational principles in dynamics and quantum theory*. Courier Corporation, 1979.

The manufacturer's authorised representative in the EU is Springer Nature Customer Service Centre GmbH, Europaplatz 3, 69115 Heidelberg, Germany. If you have any concerns regarding our products, please contact ProductSafety@springernature.com

Printed and bound by CPI Group (UK) Ltd, Croydon, CR0 4YY
26/03/2026
02078939-0001